T0220571

INTRODUCTION TO SOCIAL MEDIA MARKETING

A GUIDE FOR ABSOLUTE BEGINNERS

Todd Kelsey

Apress®

Introduction to Social Media Marketing: A Guide for Absolute Beginners

Todd Kelsey
Wheaton, Illinois, USA

ISBN-13 (pbk): 978-1-4842-2853-1   ISBN-13 (electronic): 978-1-4842-2854-8
DOI 10.1007/978-1-4842-2854-8

Library of Congress Control Number: 2017945370

Managing Director: Welmoed Spahr
Editorial Director: Todd Green
Acquisitions Editor: Susan McDermott
Development Editor: Laura Berendson
Technical Reviewer: Brandon Lyon
Coordinating Editor: Rita Fernando
Copy Editor: Kezia Endsley
Cover: eStudio Calamar

Distributed to the book trade worldwide by Springer Science+Business Media New York, 233 Spring Street, 6th Floor, New York, NY 10013. Phone 1-800-SPRINGER, fax (201) 348-4505, e-mail orders-ny@springer-sbm.com, or visit www.springeronline.com. Apress Media, LLC is a California LLC and the sole member (owner) is Springer Science + Business Media Finance Inc (SSBM Finance Inc). SSBM Finance Inc is a **Delaware** corporation.

For information on translations, please e-mail rights@apress.com, or visit http://www.apress.com/rights-permissions.

Apress titles may be purchased in bulk for academic, corporate, or promotional use. eBook versions and licenses are also available for most titles. For more information, reference our Print and eBook Bulk Sales web page at http://www.apress.com/bulk-sales.

Any source code or other supplementary material referenced by the author in this book is available to readers on GitHub via the book's product page, located at www.apress.com/9781484228531. For more detailed information, please visit http://www.apress.com/source-code.

Printed on acid-free paper

Apress Business: The Unbiased Source of Business Information

Apress business books provide essential information and practical advice, each written for practitioners by recognized experts. Busy managers and professionals in all areas of the business world—and at all levels of technical sophistication—look to our books for the actionable ideas and tools they need to solve problems, update and enhance their professional skills, make their work lives easier, and capitalize on opportunity.

Whatever the topic on the business spectrum—entrepreneurship, finance, sales, marketing, management, regulation, information technology, among others—Apress has been praised for providing the objective information and unbiased advice you need to excel in your daily work life. Our authors have no axes to grind; they understand they have one job only—to deliver up-to-date, accurate information simply, concisely, and with deep insight that addresses the real needs of our readers.

It is increasingly hard to find information—whether in the news media, on the Internet, and now all too often in books—that is even-handed and has your best interests at heart. We therefore hope that you enjoy this book, which has been carefully crafted to meet our standards of quality and unbiased coverage.

We are always interested in your feedback or ideas for new titles. Perhaps you'd even like to write a book yourself. Whatever the case, reach out to us at editorial@apress.com and an editor will respond swiftly. Incidentally, at the back of this book, you will find a list of useful related titles. Please visit us at www.apress.com to sign up for newsletters and discounts on future purchases.

—*The Apress Business Team*

Contents

About the Author

Todd Kelsey, PhD, is an author and educator whose publishing credits include several books for helping people learn more about technology. He has appeared on television as a featured expert and has worked with a wide variety of corporations and non-profit organizations. He is currently an Assistant Professor of Marketing at Benedictine University in Lisle, IL (www.ben.edu).

***

Here's a picture of one of the things I like to do when I'm not doing digital marketing—grow sunflowers! (And measure them. Now there's some analytics for you!)

I've worked professionally in digital marketing for some time now, and I've also authored books on related topics. You're welcome to look me up on LinkedIn, and you're also welcome to invite me to connect: http://linkedin.com/in/tekelsey

About the Technical Reviewer

Brandon Lyon is an expert in SEO, SEM, and Social Media and Web analytics, and is President of Eagle Digital Marketing (eagledigital-marketing.com), a full-service agency in the Chicago area. When he isn't advising local business owners and CEOs of mid-sized companies, he enjoys hockey and doing his best to survive the occasional subzero temperatures. Brandon enjoys helping companies face the challenges of the future with optimism, including navigating the treacherous waters of the Amazon ecommerce river, and taking advantage of the gold-mine in marketing automation.

Introduction

Welcome to social media marketing!

The purpose of this book is to provide a simple, focused introduction to social media marketing—for employees who may be working at a company or non-profit organization, for students at a university, or for self-paced learners. The approach is the same that I've taken in most of the books I've written, which is conversational, friendly, with an attempt to make things fun.

The experiment is to find a way to help people get started with social media marketing in a way that is fun and helps build skills—maybe through an internship, paid work, volunteer work, freelance work, or any other type of work. So the focus is on skills and approaches that can be immediately useful to a business or non-profit organization. I'm not going to try to cover everything, but just on the things that I think are the most helpful.

The other goal is to help you leave any intimidation you have in the dust. I used to be intimidated by marketing, and now look at me. I'm a marketing strategist and an assistant professor of marketing! But I remember the intimidation, so part of my approach is to try to encourage any reader who may feel uncertain about the field.

The fact is, social media marketing has a lot of options, especially in the "tools" area, and it has grown rapidly. That means there's a lot of material out there. It can be overwhelming! But it can also be very doable if you leave intimidation in the dust, take incremental steps, try things out, and build your confidence.

For example, I had a friend who used to be a journalist. He was looking for new career opportunities, so I helped get him started in social media. One of the first things he ran into was feeling overwhelmed by all the options, including all the articles about all the options. "There are so many tools out there," he used to say, "How am I ever going to learn all of them?!?"

The answer is that you don't need to learn all of them. No one can. The thing to do is to focus on some of the tools and skills and then build from there.

I encouraged my friend not to worry about trying to learn everything, but instead to just learn the basics.

My friend worked in Facebook advertising, learned a bit about Twitter, and was able to find a local agency that gave him a shot at doing some freelance social media work. The career didn't just develop for him—he had to put effort into it.

But a few years later, he's doing full-time freelance work in social media marketing and making Google ads. He was able to leave intimidation in the dust and I believe he's had some fun with it too.

This book mentions what I call the core areas of digital marketing: Content, AdWords, Social, and Analytics (CASA for short). My goal is to reinforce how all these areas are connected. AdWords is Google's tool for creating ads for search engine marketing. The inspiration came from my professional background, as well as looking at trends in the marketplace.

The Core Areas of Digital Marketing

C Content/SEO: search engine optimization is the process of attempting to boost your rank on Google so that you get higher up in search rankings when people type in particular keywords. Higher in search rankings = more clicks. The top way to boost rank is to add quality content that is relevant for your audience.

A Adwords: the process of creating and managing ads on Google (Adwords), where you attempt to get people to click on your ads when they type particular keywords in Google. You pay when someone clicks.

S Social Media Marketing: the process of creating and managing a presence on social media, including making posts, as well as creating advertisements. The main platforms are Facebook, Twitter, and YouTube, as well as Instagram and Pinterest

A Analytics (Web visitors): You can gain valuable insights when you measure the performance of your websites and advertising campaigns. Google Analytics allows you to see how many people visit your site, where they come from and what they do.

Best wishes in learning social media marketing!

The Basics of Having Fun and Building Skills

This chapter covers the basics of social media. With any luck, you'll have fun along the way as well!

Return on Investment (ROI)

ROI, or *return on investment,* is one of the themes following you around throughout this book.

But don't be alarmed.

Actually, *do* be alarmed—not by the concept of ROI, but by the consequences of *not* measuring ROI.

In the context of social media marketing, it simply measures the financial return when a business, organization, or individual spends money on social media marketing.

© Todd Kelsey 2017

T. Kelsey, *Introduction to Social Media Marketing,* DOI 10.1007/978-1-4842-2854-8_1

This issue has gotten out of control in social media—there is a real lack of measuring the true impact. Social media came along, launched like a rocket, and suddenly every business had to have it in one form or another.

Businesses became mesmerized, helped along by the media hype, and encourage by agencies that suddenly had a new source of income.

In no time at all, everyone was marching in step.

The problem is that no one really knew how to measure ROI. So people got hired—agencies got hired—but then in some cases, both people and agencies got *fired*, as businesses and organizations started asking the question:

"Okay, but how is this impacting our revenue?"

Hype: "No, it's cool man, social media is all about engagement."

Business: "All right, what is engagement?"

Hype: "Oh, it's really cool. You go out on social media, and it's all about the customer. It's like customer-centric man, you meet people in multiple touch points and engage. It's awesome."

Business: "Okay, so how does it make us money?"

Hype: "Yeah, the opportunities are tremendous! You just go out, and like make a Facebook page, and go on Twitter, and YouTube, and everywhere you can. And then you engage people with a positive brand experience."

Business: "What's a positive brand experience? Does that elevate purchase intent?"

Hype: "Totally man! Yeah, that's it. It elevates purchase intent."

Business: "Okay, cool. So how much revenue did we generate when we paid you to engage our audience and elevate their purchase intent?"

Hype: "Wow, man, that's feeling kind of black and white. You need to look at the situation holistically. People are spending a lot of time on social media, so that's where you need to be, man."

Business: "I see. So people actually make purchases on social media?"

Hype: "Well, um . . . sometimes, I think. But wait, here's this really interesting infographic that shows all the social networks people spend time on. Isn't it great?"

Business: "I suppose. So what does 'social influence' mean?"

Hype: "Yeah, I almost forgot! It's totally all about influence. Yeah, go out and identify the influencers. Make sure you have influence. It's all about influence and the power of social referral."

Business: "You mean word of mouth marketing?"

Hype: "Yeah, totally, word of mouth marketing. Like referring things you like to your peers. Bragging rights, show and tell, look what I've got, look what I found."

Business: "Okay, so how does that translate into actually making money?"

Hype: "Well, some businesses are starting to make money . . ."

So you kind of get the picture.

The Skill of Social Media Marketing

The bottom line is that regardless of how closely people are tracking the return on investment of social media, it's here to stay. And it's also true that *social media marketing*, as a core skill in digital marketing, is a top skill that gets people hired.

According to LinkedIn, social media marketing was the hottest skill that got people hired in 2013 (See https://blog.linkedin.com/2013/12/18/the-25-hottest-skills-that-got-people-hired-in-2013.)

Each year the way they refer to digital marketing seems to change, but since 2013, digital marketing (of which social media marketing is a core part) has been at the top. Demand will fluctuate over time, but we are talking about the top skills *in any field* to get people hired.

2014: https://blog.linkedin.com/2014/12/17/the-25-hottest-skills-that-got-people-hired-in-2014

2015: https://blog.linkedin.com/2016/01/12/the-25-skills-that-can-get-you-hired-in-2016

2016: https://blog.linkedin.com/2016/10/20/top-skills-2016-week-of-learning-linkedin

One of the other things I've seen in my career, which I try to reinforce in my books and in my classes, is the way that the core areas of digital marketing are related. For example, I consider social media marketing to be tightly connected to search engine optimization, or SEO. In SEO, a core goal to get higher search engine rankings is to develop and deploy content to your web site, which is relevant to your audience, relevant enough to share with other people. And where are people going to share it? On social media. And if you publish an article or blog post on your web site, where is a good place to share it? On social media.

So all I'm saying is, just like any other skill, social media marketing needs to contribute to a business or organization in a concrete way. So ROI is a really good idea to keep in mind—something to think about and learn about.

2B or not 2B? B2B and B2C

I tend to start drowning when I start swimming in acronyms, but with the right technique, you can use them well. If acronyms make you feel nauseous at all, now is the time to take some Dramamine.

If you haven't heard already, B2B and B2C are two important acronyms in business, and they have a special implication for the type of social media you might work on.

- B2C stands for *business to consumer*. So can be a retail business, such as Walmart, or Amazon, where the primary customer is an individual. You can think of B2C as a "shopper-oriented business."
- B2B, on the other hand, stands for *business to business*. This is one business selling products or services to another.

There are similarities in the way you market products to businesses and consumers, but there are also differences. In general, a business to business, B2B, approach has a longer sales cycle.

For example, if you worked at a software company and your job was to fill out TPS reports, you might suggest using an outside service to help fill out the reports. You might go online, do some research, and determine the costs. You could go to a TPS provider directly, but you could also hire an agency with expertise to advise you. So you might have meetings and submit your proposal internally. Your boss might say, "Great idea, looks like it costs a lot; let's look at this next quarter." In three months, you revisit the research, download some whitepapers, and resume the process. Then finally, you hire a TPS agency.

It's not always the case, but often there's more content related to someone buying a product or service for their business. For example, if you end up offering freelance social media marketing services, or you work for an agency or company, you might be tasked with evaluating the new tools on the market. In this case, you're in a B2B situation; you are looking at material, evaluating it, trying things out, and looking for reviews.

In a B2C situation, people also do research, but it's sometimes not as intense, and it might not take as long. For example, if you were searching for cell phones, you might look at prices, or go to a store, read a review, and make your purchase in a few weeks.

From a social media perspective, it's worth considering the context in which you're doing the marketing. As you learn about the various tools and techniques, consider whether you are trying to market something to a consumer or to another business.

One area where this concept has the most impact on social media might be in content—that is, content that you post to social media. For example, an informational article on your product might be more in-depth if it is business-oriented, and there might be more articles to peruse.

Again, it all depends on the particular product. You can think of B2C and B2B as two "roads" through social media, and the road you take affects how you approach things.

Engagement/PR

Engagement is a term you may have heard of, even if it was only in the discussion earlier in this chapter. It existed before social media, but it is *especially* present in social media.

In theory, with social media, you want to "engage" your audience through promotions, contests, content, articles, video, etc. Whatever it takes to get people talking to you and with you and talking to other people. It's a philosophy of marketing. It has its merits and limitations.

Without breaking too many rules or principles, I'm going to go out on a limb and claim that engagement as a principle probably owes more to public relations than to anything else.

The function of public relations, traditionally, has been to generate buzz about a product. For example, you might dream of holding a "publicity stunt," which draws a lot of attention and is tied to the release of an album or product.

Conversely, you may have heard the phrase, "public relations nightmare," when something goes wrong.

So a PR department's role is often to engage through whatever media channels they can, and the more awareness, the better.

The engagement side of social media might be thought of similarly—the goal is to get attention. But the question you still have to ask yourself is, "What is the impact on the business?"

If you're drawing people in with some kind of hook, and they are engaged, does it translate into sales?

Maybe, maybe not.

With all the talk of ROI in this chapter, it's fair to say that in many cases, social media could be thought of, and perhaps should be thought of, as public relations. PR is often a sunk cost where you drive awareness and manage the reputation of a company. In fairness, because so many people are on social media, it's definitely true that it's only a matter of hours, or even minutes, before some negative incident becomes a true "public relations nightmare" for any business.

It's also true that keeping your ear to the ground to detect issues early is a really good idea. You might keep that point in mind when you're reading the "Monitoring" section later in this chapter. Social media is not just for posting; it's also for listening (or reading what others post about your company).

When you're a business owner, large or small, you must simply make sure that you are out there listening to the social media world and are attentive to anything someone says about your business. This applies to local, national, and global companies. The stakes are high for any company, so it pays to listen.

Word of Mouth (WOM)

You might say that social media is a platform ideally suited for "word of mouth" marketing. The question is, can you really "make" someone share a product or idea, or can you only make it easier for them to do so?

For example, Apple Computer spends money on marketing. But their focus is on making their products great in the first place. The excellence of their products fights half the battle of marketing, and perhaps even more. Because they (mostly) produce high quality products that users love, people talk about them and tell others about them. This is self-propelled, word of mouth, marketing at its best.

So what's a marketer to do, faced with a product that is just okay, and the need to do "word of mouth" marketing? The short answer is that you can't do something with nothing.

In terms of helping a business, one of the other things that "social listening" can provide is insight into what people think about your products. It's also true that the best, most successful companies listen to their customers very carefully. One of the functions a social media marketer can provide is helping a business make and keep their products excellent. When you do that, you get that beautiful self-propelled, word of mouth marketing.

Monitoring

Some people separate social media marketing and social media monitoring, but I'm putting them under the same umbrella. Basically, there are a number of tools out there that help you listen to what people are saying.

For example, one of the tools that's often crucial to social media marketing job descriptions is Radian6, which is now part of Salesforce.

Radian6 it has all kinds of nifty ways to sift through social media and visualize the results.

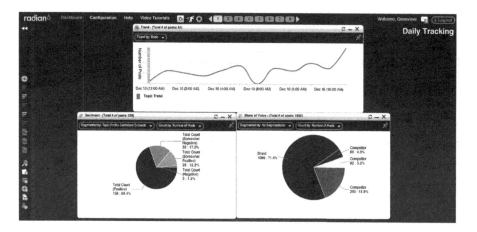

Radian6 will do things like tell you, automatically, whether people are feeling good or bad about your brand, products, and keywords. There's an automated side to social media monitoring, and it has some value, but it also has its limits. In general, social media monitoring allows you to "scan" social media and see what people are saying, down to the level of individual posts.

Entire books have been written about social monitoring, and we'll talk about it later in the book, but the bottom line is that you get what you pay for. Radian6 is pricey. On the other end of the spectrum, there are free tools. And there are hundreds in between.

As an experiment, try going to socialmention.com and typing a company name or famous person's name to see what happens.

Posting: Promotions/Content

Aside from some of the general principles, acronyms, and tools, when it comes down to it, social media marketing is about posting.

For example, here is a "promotional" post. This is an example of one kind of typical business-type post. These include coupons, discounts, special sweepstakes, etc.

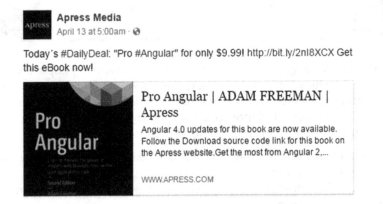

If a person is looking at the particular media channel and happen to notice the post, and then if the offer is compelling, only then *might* they click on it. If they really like the offer, they might also share it, and that's the holy grail of social media marketing—getting people to share what you've shared. It's a big game of show and tell basically.

The other kind of post is content or general. These posts are usually about something interesting, maybe relevant to the company or product, or maybe just the latest company news.

Springer Business, Economics and Statistics
April 19 at 7:38am · 🌐

Diversity empowers science and innovation http://bit.ly/2oL5pPa
#GlobalScience

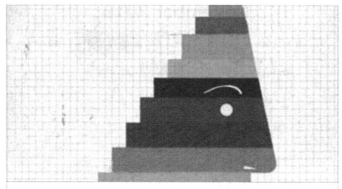

Diversity in Science: Why It Is Essential for Excellence

Science and technology are society's main engines of prosperity. Who gets to drive them?

WWW.SCIENTIFICAMERICAN.COM

The nature of your business impacts the type of posts you make. If you remember the discussion about B2C and B2B, for example, this is where the rubber hits the road.

If you have people following you on social media, your followers are business people, and you are a social media marketing agency, then your followers are probably interested in reading articles about social media. In this case, you might create blog posts that talk about social media—you might post related news or you might write a longer "whitepaper" and post that. B2B often involves longer, more in-depth content, and that content is good for posting to social media.

In a B2C setting, it's sometimes a little harder to develop content that would be compelling and interesting to your audience, but it's certainly possible. Articles, reviews, product introductions—there's a lot of variety out there.

Channels: Facebook, Twitter, and TBD

There are many "channels" in social media, but to keep this discussion focused, the book concentrates on Facebook, Twitter, and YouTube, as well as LinkedIn primarily for B2B purposes.

This graphic shows visits to different networks.

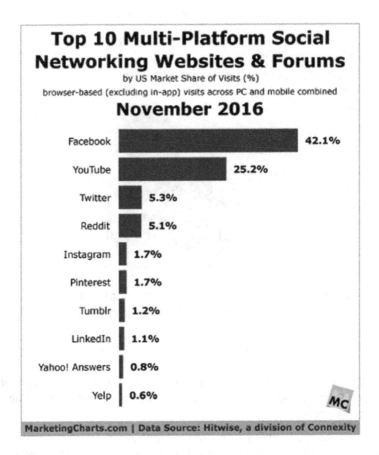

(Source: http://www.marketingcharts.com/updates/top-10-multi-platform-social-networking-websites-forums-october-2016-2-73213/.)

In general, most companies doing social media marketing of some kind will generally have a Facebook page and a Twitter account. YouTube is also popular when the business has the resources to create and post videos.

There are always new networks on the rise—Pinterest for example. One general rule of thumb is to learn about the primary networks, and then—depending on the company you work for or the company you provide your services to—learn about the audience and what types of networks the audience uses. Again, whether it's B2C or B2B could make a difference.

A consulting company selling services to other companies might have a sizable presence on LinkedIn, for example, whereas a fashion retailer might have a real presence on Pinterest.

If you're not familiar with any of the networks in this diagram, I suggest pausing, going on google.com, typing their names in, and either going to the main sites or reading Wikipedia articles about them.

Internal and Third-Party Tools

In many cases, the tools for social media marketing are built into the social network itself. For example, to advertise on Facebook, you can go to https://www.facebook.com/business/products/ads.

There are also a lot of external tools available from web sites with a free or paid service that you can use to manage some aspect of social media marketing.

For example, Chapter 7 covers Hootsuite, which can help you with a variety of social media tasks, including taking a single blog post and automatically posting it to multiple networks. It can be a real timesaver, and it's free!

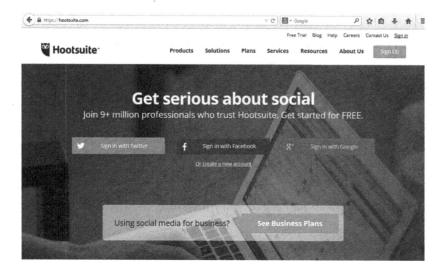

Things Change

Before I wrap up this chapter, the main point to remember is that things are always changing. Facebook, Twitter, and YouTube are probably here to stay, but things will come up out of nowhere, such as Instagram and Snapchat, which will be worth billions of dollars because so many people are using them.

It's a safe bet to become familiar with Facebook, Twitter, and YouTube, and to assume that it's a good idea to keep up with what's coming up and what younger generations are starting to use.

For example, people are using their mobile devices so much that mobile apps are drawing traffic away from web sites like Facebook. This has led Facebook to buy apps (such as Instagram) for billions of dollars and then try to figure out how to advertise on this channel.

As a medium-term, long-term prediction, more and more social media marketing will become mobile advertising. Some of these principles will remain the same, but some may change.

The bottom line is, get familiar with the big channels, but also "learn how to learn." You don't have to know everything, but you should learn about some of the growing networks.

One easy way to learn about such things is to go on sites like www.mashable.com once in a while and read the latest news. Or Google "social media news" and see what happens.

Just remember not to get buried in acronyms or buzzwords.

And always look for the ROI!

Search Drill: Learning How to Learn

As a test of the emergency "learning how to learn" system, I invite you, especially if you're using this book in a class, to take a moment, scan through the chapter if you have to, and search for more information on a particular topic, tool, or concept that was interesting to you. Any topic of your choice. Just try it out and learn something new!

Learning More

Here are some basics about Facebook, if you haven't tried it. Don't be ashamed. It's okay!

To create an account, visit https://www.facebook.com/help/345121355 559712/.

To post and share, visit https://www.facebook.com/help/33314016 0100643/.

Courageous explorers might like to explore this site: https://www.facebook.com/business/products/ads.

Conclusion

ROI! ROI! Remember ROI!

I like to be visual, so in conclusion, I present this piece of art to represent the word "fuzzy":

Because I want to make a point and claim that sometimes social media feels fuzzy to me. You end up learning about it, and it's easy to get carried away, or overwhelmed, by all the options and the hype. In many cases, there's not a very clear conversation about how it all makes money—necessary to keep you employed and to help the business in a meaningful way. It just gets kind of "fuzzy" sometimes.

Just remember, a business has to make money in order to pay you.

Skillbox: Content

This chapter takes a look at content related to social media, including tools you can use to "do it yourself". You can think of it as a toolbox of skills, or a *skillbox*. You'll take a tour of some of the things you should try when working with social media.

In some cases, if you're working for a company, they will already have systems and content, as well as any number of sources of material, for posting to social media. Still, there might be the occasional need to develop more, outside of the regular systems. During a learning phase, it can be helpful to know how to do it yourself so you can get the feel of it. You might even want to start straight out making "real" content as you're learning.

The goal is to introduce some concepts and tools that I think are worth trying, including for having material to work with when you take a closer look at social media channels later in the book.

Curation vs. Creation vs. Collaboration

The general approach I'm recommending to content, especially if you're just learning about social media marketing, is to think about one of these three areas:

- *Curation:* When you curate content, you are going out and finding it, and then sharing, reviewing, commenting on it. You might just be gathering and collecting. Even if you don't feel like you have a creative bone in your body, you can go out and find material that might be interesting and relevant to your company. It's a good way to get started.

© Todd Kelsey 2017

T. Kelsey, *Introduction to Social Media Marketing*, DOI 10.1007/978-1-4842-2854-8_2

For example, for a blog post, you might find a collection of articles on a topic and write brief summaries/reviews of them. You might provide links to the original articles and then offer some kind of conclusion. You might find YouTube videos that relate to a particular topic and either link to them or embed them in a blog post. Then you post your blog to social media and generate some traffic and awareness. You can post text, images, and videos directly to social media, but the general point is to get some awareness about your site or blog.

- *Creation:* This can be the most fun. It takes more time, but it often results in the highest quality. Writing an article or blog post or making a video—these are the best kinds of content to share on social media, because they are unique, and you can tailor them to the audience. What kind of content you create depends on the company or organization, but start by asking yourself who the audience is and consider what kinds of things they'd be interested in. You can ask people on social media or directly. It never hurts to test your ideas and find out what people are interested in, by going directly to them in some way.

  Even if you aren't a media professional, it can be helpful to at least try developing some basic content. You can always ask others to review your writing or a video you make. Trying it out might help you be in a better position to "source" it. If you end up having the choice, it can be better to focus on what you do the best and hire someone to do the rest.

- *Collaboration:* I think this one is helpful to remember, especially for freelancers, independent business owners, and students. It can be a way to save money and pool resources. Find people you can work with. It might be that you find someone who is a writer, offer to co-write some content or do some research, and get the benefit of their writing skills or their reputation. If you want to make a video, you might find people who are interested in the same outcome. You raise awareness of a particular topic or find people who want to try something new just like yourself.

  In other words, don't rule out anything, even if you don't feel confident doing it yourself, can't afford to hire anyone, and don't know where to start. You might be able to find people who are looking to collaborate, even if it's just for the learning experience.

In a business sense, it could be the same thing. If a single local business owner doesn't have enough money for a particular project, you might be able to find a collection of business owners who want to pool resources for some kind of project (such as a video featuring a local businesses). Or, using the Internet, you might be able to find similar businesses, in different areas, where you could develop an article, or video, or content that could be "re-purposed" for each business, so that with slight changes, the material could be reused.

Create a Google Account/Gmail Address

Google has a lot of free tools that make it easier to work with content, and when you have a Google account, it just makes it easier to sign in to all the tools.

As a first step, I recommend creating a free Google account by going to http://mail.google.com and clicking Create Account.

Start a Blog

There are a variety of platforms for blogging, but blogger.com is one of the easiest to use.

If you're interested in developing social media marketing as a skill, my general suggestion is to create a blog and set the goal of posting to it at least once a month (or more often). Choose a topic or tool you're learning about, or a technique you're interested in, do some research, gather some links, and get in the habit of developing some ongoing posts. It will help keep your skills current (including research), and it will also be something you can point to when you're trying to get clients or find work.

Even if you already have a blog, or have one that's been untouched in a while (raises hand), I still suggest creating a new one. It's always helpful to be able to learn new tools. Another reason it's helpful is because you could end up in a situation where a client might want to create a blog, and you can help them get started, by being familiar with and showing them different tools.

This same principle applies to some of the other tools we're taking a look at in this chapter. I recommend trying them in some way. This is good for building your own skillset, but can also help you be able to show a client (or potential employer) someday.

To get started, visit http://www.blogger.com and either sign in with your Google account or click the Create an Account link at the bottom.

Then on the blogger site, click the New Blog button.

For practice, don't be too concerned with the title. You can change it later easily, and you can also create/delete blogs easily. Feel free to try "Social Media Perspective" as a title.

Blogs List › Create a new blog

Title	Social Media Perspective
Address	.blogspot.com

You can add a custom domain later.

The title is simply what appears visually at the top of the blog. The Address is the opportunity Google gives you to create a custom address. Because it's a free tool, you might have to experiment a bit until you find one that's available. Type in ideas in the Address field, and see what happens:

Sorry, this blog address is not available.

What you're doing is coming up with the custom portion of the blog's address.

socialbuzznews| .blogspot.com ✓

This blog address is available.

It turns out for this example, the address socialbuzznews.blogspot.com is available. The link for this blog would be http://socialbuzznews. blogspot.com.

After you choose a title and address, you can choose a template for the look and feel of the blog, which you can also change later:

After you've selected a template (I recommend starting with Simple), click the Create Blog! button. Using these simple steps, you've created a blog and can start blogging!

Your mission if you should choose to accept it is to make a sample post, and then share the link on Facebook or via e-mail with someone.

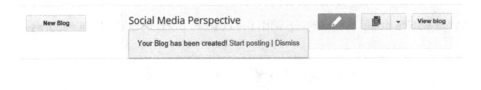

Note   One way to "cheat" if you forget the blog's address is to click on the View Blog button (see screenshot above), which will open the blog in your browser. Then you can copy the link from the address field and paste it into Facebook or an e-mail.

To learn more about Blogger, access the Settings menu (click the little gear icon) when you're signed in to Blogger and select Blogger Help:

There are a variety of helpful articles:

You can always go directly to the help center with this link: `https://support.google.com/blogger`.

Search Drill: Find a Blog

I recommend that you try creating a blog (and even setting a reminder to make a post once a month or more).

I also recommend trying to find a blog that you'd be interested in reading, as an example of "curating" content. Find it on Google by searching for something like "social media blogs". Find an article that looks interesting and make a note of the link. When you create a Facebook page, you may end up wanting to post that link on your page. Or, you might want to link to someone else's blog when you are writing about a topic.

Create a Free Web Site

Whether you are developing your own social media presence or working on someone else's, it can be helpful to consider making an "official" web site for a project, event, campaign, or client. A blog technically is a web site, and blogging platforms such as WordPress (www.wordpress.com), which has free and paid versions, have grown to the point where they can serve as fully-functional web sites, depending on how you organize them.

It can be simpler at first to think of a blog as a place you post "ongoing" content, such as a library of articles, and your "main site" is the reference material that may not change as often, whether it's for a business or organization.

In a company setting, knowing how to easily create separate web sites can also be part of social marketing, in which case you create *microsites*. For example, a company might have a special promotion that is displayed on social media and Google ads. You might be able to put it on the main official web site, but there might also be reasons where a separate microsite is a better idea. For example, it might be easier than trying to connect with their system, or have to work with their IT people, etc. Microsites are used often for special promotions or offers.

I think Google Sites is a good tool for anyone creating a simple web site, and more tools are mentioned at the end of this chapter. I recommend you try making a Google site, and keep it as part of your arsenal. You might even want to have a Google site be your main site, such as your freelance business, etc.

Note   With Blogger and Google Sites, even though you choose a custom "long" address provided by Google, you can also use your own web site name. Web site hosting companies will sell you web site names, or web "hosting" space, where a traditional site can be built using HTML, etc. Another advantage of Google Sites and Blogger is that they're entirely free, but you can still point your web site name to them. A web site name like www.toddkelsey.com may cost only

$10 a year, kind of like a copyright, whereas a web site hosting account, to do a manually-created web site, might start at $10/month. What I noticed over time is that for some simple sites, I personally prefer Google Sites because it allows me to easily post content, there is no monthly cost, and it is basically more sustainable.

And having your own web site name comes across slightly more professionally – so you might want to file that away. You can look for names on web sites like www.godaddy.com, and start an account (a web site name such as www.toddkelsey.com is also known as a "domain" name). And then you can "point" your web site name to your blogger blog or Google site. If you end up wanting to try that, I'd suggest looking into the help sections on blogger and Google sites about "web addresses". Best wishes!

To get started, go to http://sites.google.com. Log in if you need to or click the Create Account button.

Like Blogger, Google Sites packs a lot of power. The other advantage of sites like Blogger and Google Sites is if you're not a developer, you don't need to have technical skills in order to create web sites using these tools. It might be a good alternative where you provide the service of focusing on the content for the site or marketing and can whip up a site, without necessarily having to hire a web developer. There are limits, of course, but it can also be a starting point. For example, use Google Sites to gather content to begin with. Then organize and prototype, and finally, when you have a better idea of where things are going, hire a designer/web developer.

I think the same applies to marketing microsites. Say you develop a social media campaign and want to have a microsite. A large part of the battle is just developing the content, gathering it in the first place. You can start with a free tool, take it to the limits, and then decide if you need to have a more flexible or professional looking design.

■ **Note**   One other thing I'd say is that with the rise of mobile devices, the rules of design are changing a bit. It's not that you should ignore design, it's just that you can't fit as many visuals on a smaller screen, and the mobile user may be more interested in just getting to information. In other words, designing a site that looks good in a mobile setting (usually this means its simpler) is not such a bad idea.

So, back to Google Sites.

Remember how things change in social media? They change on Google too. Google experiments and improves its products and it has been experimenting with a "New" and "Classic" version of Google sites. You may see a button like this:

This chapter deals with the Classic version, but the principles are the same. It's that easy—just click the Create button.

(If you're taking it for granted how easy it is, try going to Godaddy.com and looking at how much effort is required to start a hosting account and get a web site started, with a content management system like Drupal, a web site builder, or even using a manual tool like Dreamweaver. I guarantee after trying that approach, you'll appreciate how much time you're saving by just being able to click Create. Thanks, Google.)

Just like with Blogger, Google Sites has pre-built templates you can choose from. In Google Sites it is a little more tricky to go back and change things later, so until you explore how to customize your site, I recommend choosing the Blank template.

You select the template and then choose an address, just like you do in Blogger.

You click in the Name Your Site field and type a name, which is like a title and can be changed easily later. Then you'll have to experiment and try different "site location" names. You can also click on Select a Theme or More Options, but I suggest keeping it simple initially. (In Google Sites, the "themes" are what you can come back and easily change later, and they provide some basic customization in look and feel.)

Name your site:

> **socialmicrosite**

Site location - URLs can only use the following characters: A-Z,a-z,0-9

https://sites.google.com/site/ | socialmicrosite

▸ Select a theme

▸ More options

Type the code shown:

CREATE

You also have to type in the CAPTCHA code (such as "bitsu" shown in the figure) before you can click the Create button.

As you're typing in site location names, Google may tell you that the one you want isn't available. If so, you might have to keep trying:

Site location - URLs can only use the following characters: A-Z,a-z,0-9

https://sites.google.com/site/ | mysite | ✕

The location you have chosen is not available. Learn more...

And then—voila! You have a new web site.

To learn more about Google Sites, go to either of these links, which point to the same place: https://support.google.com/sites/?hl=en#topic=1689606 or http://tinyurl.com/googsite-help.

My basic recommendation is to make a site that highlights your portfolio and work experience. You can also make a practice site for a potential client, such as for an imaginary local business or a promotional campaign of some kind.

Other Systems for Making Web Sites

There are several more popular options that include free/paid options for making web sites:

- http://wix.com
- http://weebly.com

Take a look and try them out. Even if you're an intern at a large corporation, sometimes big companies use microsites too—especially in marketing situations.

Make/Edit a Video

It's definitely true that entire books, or sets of books, could be written about each individual section of this chapter, but I'm recommending some simple starting points that are helpful to explore. With video, as with blogs and web sites, you may end up wanting to hire an agency or professional to make a video as part of a campaign. At the same time, individuals can do pretty interesting things on their own with something as simple as an iPhone.

There are a variety of ways to make videos, and if you don't have a video camera or iPhone, I recommend getting an older iPhone or just an iPod (yes, they still make them) for the purpose of filming video. You can get a used iPhone fairly cheaply, connect it to WiFi, and upload videos directly. As of this writing, even a new iPod is $200, which is not too bad, with no monthly cell phone bill. (It's also true that a used iPhone could be used for making video and some other uses, even over WiFi.) If you've always wanted an iPad, there are used options there too.

The reason I recommend one of these options for experimenting with video is because it's simple, easy to use, creates high-quality video, and you can upload the video directly to YouTube.

Getting familiar with it, including making simple videos (such as interviewing someone) and posting them on your blog or web site, can be good experiments—all good ways to build skills and your portfolio. In more professional environments, even if you're requesting a budget to hire a professional videographer, you might still want to try prototyping your ideas with cheaper modes of video.

Keep in mind that some of the most popular videos on YouTube were created with really simple equipment—it's more about the idea than the equipment.

I'm not Apple-biased, it's just simple, and that's good for beginners. Point and click and you can make simple videos. Then, you're one click away from being able to upload to YouTube:

Part of the reason I am suggesting this kind of arrangement is to keep the technical hassles to a minimum, so you focus on the content. The easier it is, the more fun it is, the fewer hassles there are, and the more confidence you will build from trying things out. Just a starting point.

By comparison, you can certainly get a digital camera with built-in video capability, or any number of dedicated cameras, load video editing software on your computer, and then transfer the video to your computer. In fact, I encourage you to explore that approach at some point.

But to begin with, I recommend getting a cheap iPhone or iPod or iPad, perhaps one that isn't the latest version. Just make sure it has built-in video.

I'm suggesting you try some videos without even editing or just by using YouTube's built-in editor. Meaning you shoot some short clips on your simple mobile device and upload them to YouTube. Then you edit them on YouTube. Simple, fairly easy, and lets you focus on the content.

Another option is the Android, an alternative to an iPod or iPhone. My general experience is that Apple takes a lot of time to provide a good, simple, fairly stable user experience, whereas with the Android platform, it depends on the manufacturer. The time you spend on figuring things out may take away from having fun. However, to save money, or to avoid Apple specifically, Android tablets are certainly a good option.

At the time of this writing, you can get an Android tablet, with the capability to shoot video and upload it to YouTube, for about $50 USD.

Editing on YouTube

My recommendation is not worry too much about the video. Just try shooting "something," even if it's really simple, and then upload it to YouTube. The YouTube Editor is located at https://www.youtube.com/editor. If you are new to Google Accounts or YouTube, you might need to set up a profile on YouTube:

How you'll appear

Add photo	John	Doe

To use a business or other name, click here.

Gender   **Birthday** *(i)*

Male ▾   April ▾   1 ▾

Cancel   Continue

Whatever YouTube videos you've uploaded to a particular account will appear, and the YouTube Editor enables you to do some basic editing online. I think it's nice, even if there are limits, to be able to try video by shooting on a mobile device, uploading directly, and then editing right there online.

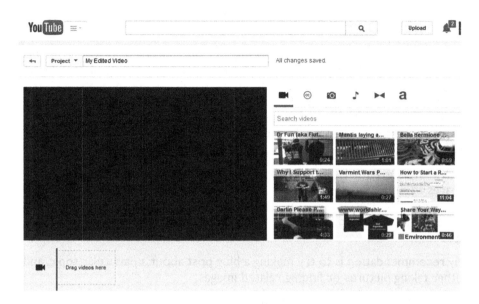

As with the other topics, entire books could be written on working with video and you may want to look at some. However, I also encourage you to experiment by making a simple short video and not worrying too much about technique yet. I think it will give you confidence. With regards to social media marketing, I think video is one of the strongest, long-term features that will still be around as social networks rise and fall. Getting familiar with making videos, even if they are rough or simple or mainly informational, is a good thing.

To learn more about the YouTube Editor and how to use YouTube, visit https://support.google.com/youtube.

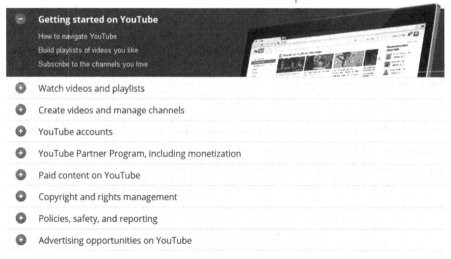

Adjusting Digital Images

Another area I recommend exploring for the purposes of social media marketing is digital images. This can mean something as simple as looking on Google for images. Once you find an image you like, you can right-click on it and save it to your computer (Windows) or hold Ctrl down and click on the image to copy it (Mac). Try some simple image searches and practice with those images first. Then explore royalty-free image collections and services like clipart.com.

The point is, eventually you'll probably want to at least try playing with adding images to a blog post or web site, and it's good to know how to work with images or pictures for posting to social media. For example, even when you have a text-based post, an interesting, related picture can lead to more attention.

My recommendation is to try making a blog post about a particular topic, and either taking pictures or finding related images.

Even though when you upload images to web sites or blogs, there are often built-in "resizing" functions, you still might want to try a tool like www.picresize.com to crop an image or resize it. (Try looking at the help section.)

In other words, there are online tools you can use, without necessarily having to learn a complex image editing program like Photoshop (costly) or Gimp (open source at www.gimp.org).

Taking Screenshots

For grins, and to add some value to your blog posts, I recommend learning how to take screenshots, especially students and interns, or anyone for that matter. Think of it as a way of taking pictures on the web. For example, most of the images in this book I'm writing are screenshots, where I'm taking a picture of something and then discussing it.

Including a picture of a piece of software or a web site is an easy way to add a visual to blocks of text. Greenshot is one free tool you can try at www.getgreenshot.org.

■ **Note** On the Mac, you can also just press Cmd+Shift+4 on your keyboard to grab a screenshot. Then you click and drag to indicate the screen area you to capture. (See http://guides. macrumors.com/Taking_Screenshots_in_Mac_OS_X.)

Your mission should you choose to accept it is to choose a feature in something like Blogger or Google Sites, or even Facebook, explore it, take a screenshot or two of it, and put them in a blog post.

Learning More

For a free web site maker, check out `http://sites.google.com`. It's all-purpose, free, and includes some capacity for customization.

A couple more options that are popular and have free/paid options include the following:

- `http://wix.com`

- `http://weebly.com`

- `http://www.shopify.com` (this is strictly not free, but includes a free trial for e-commerce sites)

Conclusion

This chapter has definitely been a whirlwind. There are a lot of loose ends. But my approach is to help you get started in a way that's simple and fun. If you've been glancing through it, I recommend going through the chapter more carefully, maybe trying one tool a day, or per week, and getting something going as an example. As you try social media marketing, you'll want to have material to practice with, so it's better if you can try making some of your own sample content to save for future examples.

Facebook Pages

The goal of this chapter is to introduce the concept of Facebook pages and walk through the process of creating one. The chapter starts off with a discussion of what a Facebook page is, as well as the pros and cons of using Facebook, and then there's a tour of creating one. If you have not created a Facebook page before, I recommend giving it a shot, even if you don't have an "official" project yet. As with many free tools these days, you can come back and delete it.

What Is a Facebook Page?

A Facebook page is a central "hub" for a business or non-profit organization to establish a social media presence on Facebook. It's basically like having a web page or web site of your own, but "within" Facebook, and following its format. As a Facebook page owner, you can make posts of various kinds, including text, pictures, video, etc., and if users click the Like button on the page, then in theory, they will receive posts in their newsfeed when they log on to Facebook. Most social media marketers consider a presence on Facebook a must-have, but it's worth knowing the limits and costs, as we'll discuss.

Common Facebook pages includes ones about authors, movies, and entertainment.

For example, if you go on Facebook and type "matrix" in the search box, you can select "The Matrix" Facebook page. At this time, it has about 7 million people who've liked it.

© Todd Kelsey 2017
T. Kelsey, *Introduction to Social Media Marketing*, DOI 10.1007/978-1-4842-2854-8_3

Facebook pages also have "web addresses," which can be used in promotional materials to advertise or link to the page. For example, the direct link to "The Matrix" Facebook page is `https://www.facebook.com/TheMatrixMovie`.

■ **Note**   When you create a Facebook page for the first time, it may be a longer link, such as `facebook.com/pages/name of your page/11887376363873`. But there is a process to choose a shorter name for the Facebook page so that the direct link is easier to remember, like the one in this section for "The Matrix".

When you access a Facebook page, depending on how things have been designed, you'll generally see a prominent image at the top, a box with the Profile icon, and an area for posts below it.

Notice the "Add a Cover" image. That's one of the options for customizing your Facebook page, which—nudge, nudge—is an opportunity to work with digital images, as described in the last chapter. When you start making profile icons or cover images, it's common to start with a picture and then resize it/ crop it to a particular size.

Here is an example of a Facebook page that I liked. Because I liked it, a post showed up in my newsfeed.

In addition to entertainment and business, in the last few years, Facebook pages have also been increasingly been built around causes, and have had a direct impact on world events, both through official pages of movements, as well as independent pages.

For example, during the Arab Spring, I created a Facebook page with some music and poetry to encourage people who were protesting to bring about democracy in their countries in the Middle East. Check out www.facebook.com/freedomsongs.

You might be interested to review/discuss the nature of how the revolution in the Middle East was sparked. It wasn't "caused" by social media, but it just so happens that a Facebook page was a central method of communication. See http://www.nytimes.com/2012/02/19/books/review/how-an-egyptian-revolution-began-on-facebook.html.

This was the page that basically started it: https://www.facebook.com/elshaheeed.co.uk.

You might also be interested in looking at a video I made, which tours the "Freedom Songs" Facebook page I made in support of the revolution, with a discussion of how Facebook advertising was used. It also includes some of the "behind the scenes" information on the Facebook page, including statistics that were gathered about people who visited and liked the page from various countries.

The title is "How to Start a Revolution (or Help One)" and a direct link is: http://tinyurl.com/fb-arabspring.

Facebook Pages: To Do or Not to Do?

In my opinion, one of the most important points to keep in mind about Facebook pages, whether you are creating one for a business or non-profit organization, are the limitations. Facebook pages are often considered a central piece of social media strategy, and the general mindset is that they are a must-have. However, some technical changes at Facebook have changed the effectiveness of pages as a promotional tool, which is worth looking at.

Principle: What Is Facebook's Business Model?

One thing that helped me in a book I read about Google (*Winning AdWords* by Danny Sullivan) was a challenge that Sullivan made about understanding Google's business model. I think a similar principle can help social media marketers look at Facebook with a critical eye.

In Google's case, the discussion was about techniques for getting a page listed on Google's search engine for free, which is also known as SEO, or Search Engine Optimization. Google allows just about any web site to be listed for free, but it also offers the opportunity for paid advertisement. (The paid advertisement is known as Search Engine Marketing, or SEM).

In terms of business models, Google doesn't technically make money from offering free placement on its search engine. When people submit links, and when Google crawls the Internet to discover sites automatically, it adds value to Google. However, the way Google makes most of its money is through ads. Tens of billions of dollars in ad revenue.

So the challenge is that, even if you put effort into the "free" techniques, it stands to reason that Google's business model is about paid advertising, and that putting resources and effort into the paid techniques is more sustainable and reliable long term. Basically, Google doesn't "promise" anything, even though an entire industry has grown around SEO. Even though the free Google placement, SEO, is a good thing to do, you can't *necessarily* rely on it.

Once in a while Google will change the way web sites appear, which can have a pretty big impact on businesses that have made assumptions about their "free" SEO campaigns. In short, understanding where a business makes money can help you plan long-term and think critically about marketing in its "channel".

In terms of social media, historically what happened is that Facebook grew, and then started thinking more about how to make money, which ultimately had a direct impact on Facebook pages.

The implicit assumption used to be that when you got people to like your Facebook page, the posts you made would appear in their newsfeeds. Kind of like a subscription. Social media was touted as a method where you could more or less get free advertising—self-promoting on "social media"—and then build a presence by putting content out there.

But then Facebook started charging for "boosting" posts.

I think the basic principle applies to Facebook too. Just keep in mind the question—how does Facebook make money? The answer is through advertising. So it's likely that while you might get something for nothing, in the end it might not be sustainable, and it's important to have realistic expectations.

With social media, it's especially important to think critically, and skeptically, because of all the hype around it. So as a social media marketer, you could have clients who know "they need to have a social media presence," and you might be able to help them based on this "must-have" attitude. Over time, I think it's important to ask the question about return on investment and understand what the impact is. You have to question assumptions that people have and help them understand the true value. In the end, look at the numbers.

Here's a small-scale example. For my "NPOEx" page, at the time of writing, there are 220 Likes.

So I might make the assumption that if I post something, in theory the people who like the page would get the post in their newsfeeds. (Even if they do, there's no guarantee that everyone reads all their posts, which is another limitation of Facebook that's important to keep in mind.)

So I make posts and hope for the best.

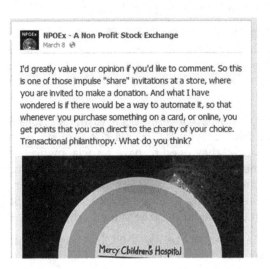

The important question to ask is how many people actually saw the post?

When you post to your page on Facebook, you can see how many people it was "served" to.

The graphic shows that the Facebook post, on a page where there were 220 Likes, was served to only 10 people. That means 10 people potentially looked at it.

The red arrow points out the relatively new Facebook "feature," which allows you to boost the post. You'll quickly see how, in order to reach more people, you have to pay Facebook.

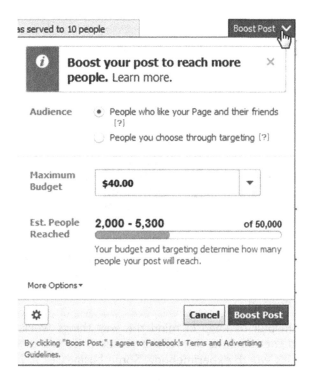

Granted, in this "boost" scheme, Facebook is including a way to reach more people beyond your audience. But it is also charging you to reach people who liked your page. So if you thought reaching your fans was free, think again.

Boosted Posts

When you share useful and engaging posts on your Facebook Page, you're building relationships with current and future customers. For as little as $5, you can boost a post so more people will see it when they visit Facebook. Just click on "Boost Post" while you're creating a post, or once it's already on your Page.

You can choose to reach more people who like your Page and their friends, or new audiences. Curious about which posts are most interesting to people on Facebook, or want to know more about the people who like your Page? You can learn more by looking at your Page insights.

Click on your Page profile photo to get started now.

From my perspective, I think it ends up being a little like Google ads. There's competition, and to a certain extent, it's like Facebook is treating people's newsfeeds like Google search results. You can pay Facebook to get a better change of being noticed.

In the end, it's helpful to keep in mind the way things work, decide what your goal is, and then look at the results to see if it's worth it. There's no rule either way; it's worth experimenting. Some businesses have had success gaining followers and leveraging Facebook pages to help establish and maintain their presence on social media. From a credibility perspective, many argue that part of maintaining "social credibility" is having a social media page.

But there are also instances where the effort and money put into a Facebook page leads to questionable results.

For example, here's an article where someone looked with a critical eye, did some testing, and found out that Facebook pages, for them, weren't such a good idea:

"Facebook Pages Are a Bad Idea" http://www.forbes.com/sites/ elandekel/2013/01/22/facebook-pages-are-a-bad-investment-for-small-businesses/

In the end, I think Facebook pages are definitely worth learning about. They are legitimate hubs for social media. I recommend thinking of Facebook as an experiment. Try it out and see how it works, then look at the numbers.

In terms of content, if you are creating content about your organization and giving it a home on your web site or blog, the most important thing is that you are building up an archive of stories about your organization. Even if you never posted them on social media, they would be important. (In part, because they help a web site get recognition on Google. In other words, even though SEO has its limitations, just putting content on your site draws people there and helps you with Google search results.)

If you already have the content, why not share it on social media? Just remember to "drill down" and look at how many people *are actually seeing it*—to set realistic expectations.

Part of your social media strategy might involve trying different ways of increasing engagement, such as creating a video that has more of a chance of being shared because of its appeal—the story, the humor, the relevance of information. The same idea applies to articles.

The bottom line is—social media is worth trying.

Creating a Facebook Page

Creating a Facebook page is fairly easy. Just go to https://www.facebook. com/pages/create to get started.

There are a variety of page types, and I recommend clicking on various ones to try them.

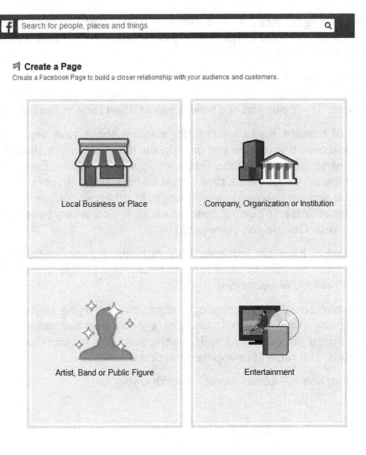

For example, you might try clicking on the Company type:

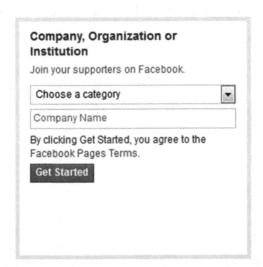

Then you choose a category:

Type in a name for the business and then click Get Started:

Next you will be led through a series of screens where you can enter basic information.

As a learning exercise, I recommend making a checklist that you can refer to later, in order to assemble this content. But for now, I suggest clicking the Skip button on each page:

Set Up Free Social Media

| 1 About | 2 Profile Picture | 3 Add to Favorites | 4 Reach More People |

Tip: Add a description and website to improve the ranking of your Page in search.
Fields marked by asterisks (*) are required.

Add a few sentences to tell people what your Page is about. This will help it show up in the right search results. You will be able to add more details later from your Page settings.

155

*Tell people what your Page is about...

Website (ex: your website, Twitter or Yelp links)

Choose a unique Facebook web address to make it easier for people to find your Page. Once this is set, it can only be changed once.
http://www.facebook.com/ Enter an address for your Page ...

Is Free Social Media a real organization, school or government? ○ Yes ○ No
This will help people find this organization, school or government more easily on Facebook.

Need Help? | Save Info | Skip

The second screen allows you to upload a picture. (Nudge nudge—see Chapter 2.)

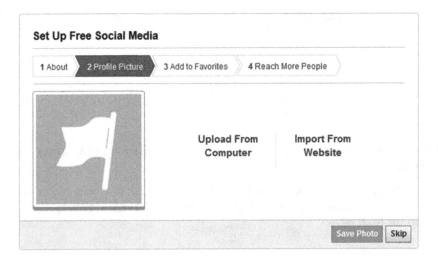

Set Up Free Social Media

| 1 About | 2 Profile Picture | 3 Add to Favorites | 4 Reach More People |

Upload From Computer   Import From Website

Save Photo | Skip

This is probably worth doing—it just makes your page easier to find and manage when you sign in to Facebook. Favorites refers to your "Favorites" area on the home page when you sign in to Facebook, until/unless Facebook changes its format:

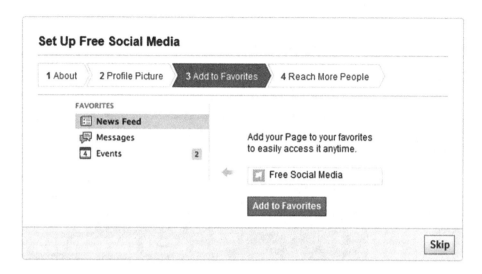

Then you get an introduction to Facebook advertising, ending with inviting you to create an ad. How does Facebook make money? Through ads.

As you'll see in Chapter 4, there's a variety of ways to create an ad. I recommend skipping this page for now.

And then, voila! You have your draft Facebook page:

Congratulations on making a Facebook page.

Remember, you can always access Facebook Help for more information: https://www.facebook.com/help/364458366957655/

This page has good, basic information about creating and managing a Facebook page, which is worth reviewing.

Posting to the Page

The general point of having a Facebook page is to get people to like it by promoting it and to make posts to it.

One point to keep in mind is that Facebook allows you to post as "yourself" to a Facebook page, but it also allows you to make posts "as" the page.

You are posting, commenting, and liking as **Todd Kelsey** — Change to NPOEx - A Non Profit Stock Exchange

For example, on the facebook.com/npoex page, I can post to it as Todd Kelsey. In general, it's better to post as the page, so that the posts are marked as coming from NPOEx.

Just something to keep in mind.

Content! Content!

Here's another opportunity to revisit and reconsider the chapter and skills about content. Here's where you are asking yourself, "Okay, what can I post?"

And here's where, if you haven't already, you may want to make a blog or post and then post that link (about your organization, or thoughts, or whatever) on Facebook.

To post, you just click in the post area. (For example, you can use the "What have you been up to?" area.)

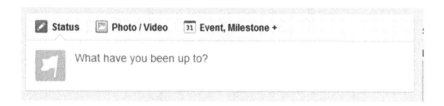

You could type something, like Hello world!:

Then you click the Post button.

You get a post like the following one:

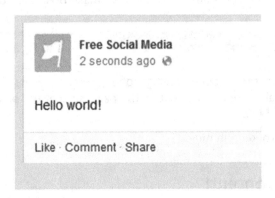

Accessing Pages

You can access your Facebook pages by going directly to them, or by logging in to Facebook, clicking on Home, and then looking for the page on the left side:

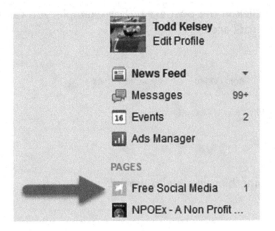

Facebook Page Links

If I access my page using the previous method above, I get a link like this one in the browser's address bar: https://www.facebook.com/pages/ Free-Social-Media/1510620519168937?ref_type=bookmark

This link can be shortened to https://www.facebook.com/pages/ Free-Social-Media/1510620519168937.

(You don't need the information beginning with the question mark.)

This is the basic link to the Facebook page. People can also search on Facebook for the page based on the name you choose:

Facebook Page Usernames

In general, at some point, you'll probably want a shorter link so your page is easier to find and promote.

Instead of a long link like https://www.facebook.com/pages/Free-Social-Media/1510620519168937?ref_type=bookmark, you'll want a shorter link, which is easier to promote, such as www.facebook.com/npoex.

In order to create a shorter link, you have to create a Facebook page username.

To get one, go to your Facebook page and click on Settings:

Then go to Page info:

Click on Enter a Facebook Web Address:

Facebook Web Address   Enter a Facebook web address

Then, click on Create a Web Address for This Page?. You can click on the question mark (?) to get more info if you need to:

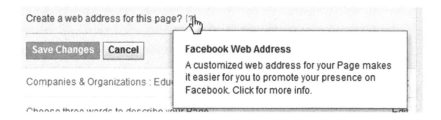

This helps you create a web address name that hasn't been taken on Facebook:

Create your Facebook web address

Easily direct someone to your Page by setting a username for it. After you set your username, you may only change it once.

Page: | Free Social Media ▾ |

Facebook Web Address: | www.facebook.com/[Choose a username] |

Check Availability

Click on the Facebook Web Address field, type a name, and then click on Check Availability. When you find one that's available, review the info and click Confirm:

Username Available

FreeSocialMediaInfo is available.

 Several things for you to remember:

- You can only change the username of Free Social Media once after you set it the first time.
- You can't transfer the ownership of a username to another party.
- You can't violate anyone else's trademark rights.
- If you are acquiring a username to sell it in the future (squatting), you will lose it.
- Usernames may be reclaimed for other unauthorized usages.

Are you sure you want to set FreeSocialMediaInfo as Free Social Media's username?

Confirm Cancel

Then click OK:

Then you will have a shorter link, such as `https://www.facebook.com/FreeSocialMediaInfo`.

Some businesses advertise by removing the `http` and www, such as `facebook.com/FreeSocialMediaInfo`.

The More Things Change...

Another point I think is helpful to keep in mind is that Facebook and other social media channels regularly change their features, often out of a desire to improve or be more secure. So the way that Facebook pages work may change (for example, there weren't originally cover images), the interface used to work with them might change, or additional features will be added.

My general suggestion is to explore Facebook and review the help. Facebook often sends e-mails with updates as well. Learn how to learn and don't be afraid to experiment as new features come out.

Learning More

Facebook's help section provides more good information at `https://www.facebook.com/help/364458366957655/`.

Conclusion

Best wishes in exploring Facebook pages! My general recommendation is to focus on the *content*, and not the channel, such as Facebook. Once you are developing content, experiment with posting it to various social media channels. Use Google to search for articles that discuss social media strategy, including what to post, how often to post, etc. A safe rule is to consider what your audience would be interested in reading.

Remember to always see how many people you are actually reaching with your posts!

Facebook Ads

This chapter is an introduction to Facebook ads, which are commonly used for social media marketing. There's some discussion of the type of ads and the traditional value of using them, as well as some things that have changed in Facebook as new types of ads have been introduced. You also have an opportunity to create a Facebook ad. You will advertise a blog or site you created after reading Chapter 2 or use Facebook ads to advertise a Facebook page that you created in Chapter 3. My recommendation is to make or find a web site that you want to advertise a bit, create an ad, and try running it for a week or so. Toward the end of this chapter, we look at "monitoring" Facebook ads, in order to evaluate their performance.

What Are Facebook Ads and Why Should You Use Them?

The most common type of Facebook ad is the kind that appears on the right side of the page when you log in to Facebook. Here are some examples.

© Todd Kelsey 2017
T. Kelsey, *Introduction to Social Media Marketing*, DOI 10.1007/978-1-4842-2854-8_4

SPONSORED 📑   Create Ad

Create Your Own App
ibuildapp.com

Free and Easy! No
Coding! Get Mobile Now!

Doctor's NightGuard
amazon.com

Doctor's NightGuard
Dental Protector is
available at Amazon and
other fine retailers

Free WebEx Account
webex.com

Meet with your colleagues
from wherever you are.
Backed by Cisco reliability
and security.

Another type of ad appears directly in your newsfeed, alongside posts from your friends and from Facebook pages that you have liked. At the very top, it says Suggested Post:

You don't need to be a business selling a product or service to use Facebook ads—anyone can use them. For example, I made this little ad to see if I could lure some people into reading a science fiction short story:

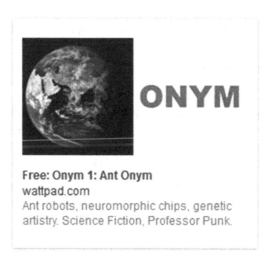

It leads to a site called Wattpad. I wasn't selling anything per se, I was just promoting some fiction that I had written:

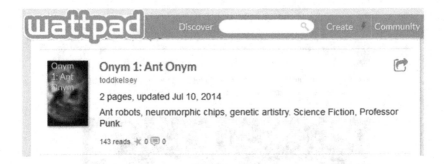

When to Use Facebook Ads

The bottom line on Facebook ads is that in the foreseeable future, people are still spending a lot of time on Facebook. There's evidence to suggest that younger people are spending more time on messaging apps like Snapchat, WhatsApp, and others, and spending less time on Facebook (since their parents are on it?), but at the end of the day, people are still spending a lot of time on Facebook.

So one principle about Facebook ads, and any kind of social media marketing, is that it "goes with the flow". That is, wherever people are spending time.

Note Another point to keep in mind about "trending" is that people are increasingly spending online time on mobile devices, and the way ads appear on mobile devices is different than "desktop" ads that appear when you're on Facebook or other social networks based on using a desktop or laptop computer. In the case of Facebook ads, you'll see that there's an option for making them available to mobile users. If I'm not mistaken I think that roughly half the ads on Facebook at the time of this writing are displaying to mobile users.

In terms of when to use Facebook ads, there's not really a right or wrong time, but the most common uses include when promoting Facebook pages, where the ad is on Facebook—it is displayed to Facebook users—and the goal is to get people to visit the Facebook page and click the Like button.

As discussed in previous chapters, one fundamental question to ask is whether you are getting a return on investment from something like a Facebook page. If you or your client has the goal of "increasing likes," then a Facebook ad is a good way to get a social media presence established. At a very basic level, even if you're not planning or necessarily assuming you'll actually make money with social media marketing, you might end up seeing it as a credibility issue—"because everyone else is doing it". It is arguable that people use company Facebook pages as a way to find or learn about a business.

What I've seen professionally is a trend toward using Facebook ads for "traditional" display advertising, meaning using an ad on Facebook to point to an external web site, with some kind of image and text. This is an example of overlapping with "traditional" Internet marketing and social media marketing. In traditional Internet marketing, a display ad might appear on a news web site, for example, and consist of some kind of image or animation and text, which points toward some kind of web site.

Social networks came along and social media marketing was born. The general goal and scope of social media marketing is to establish and maintain a presence on social media. But a Facebook ad can be used for "traditional" purposes as well—it might just be to advertise a web site.

This is a topic worth discussing or even debating and some argue that any presence on the Internet is necessarily social, and that social media marketing helps businesses understand that the new era of consumer interaction is more personal. It could be argued that a "traditional" web site is part of an overall social media presence.

Purchase Intent

I think another helpful principle when considering Facebook ads is the question of *purchase intent*. In general, placing an ad on a search engine is different than placing one on a "general" site. People are more likely to go on Google or other search engines when they are researching a product—basically when they have an intent to purchase. While it may change over time, it's generally accepted that people are mostly going on Facebook to see what their friends are up to or to post material—not necessarily to make purchases. There are posts that amount to social referrals, such as, "hey, look what I've got" or "hey, I recommend this". However, it's still the case that the purchase intent is probably significantly lower than with search engines.

This intent translates to statistics—even though there is less of a chance of a purchase when on Facebook, advertisers still place ads there. The statistics may be lower on getting purchases, but it can still happen. Just something to keep in mind.

Metrics for Measuring Facebook Ads

While learning about Facebook ads, there are some concepts that are worth considering. Facebook and other ad platforms display a lot of information when you're running an ad campaign. You don't necessarily need to know all the particulars in order to make use of them. I recommend exploring and seeing this as an experiment. The more you use Facebook ads in business settings, the more you'll be interested in some of the finer points over time.

At a high level, when you create an ad on Facebook, there will eventually be a variety of information that shows you how your ad performed:

A central concept, perhaps *the* central concept, is the question of how many clicks the ad got:

81

In the end, the purpose of the ad is to get people to click on it and to visit a particular page or site.

Another concept to consider is *click-through rate*. It's also known as CTR.

For example, in the following figure, you can see in the Reach category, that in theory, the ad was displayed to 61,432 people. There were 81 clicks. So the click-through rate was .081%.

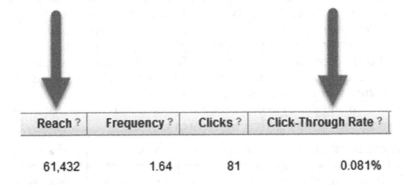

Reach ?	Frequency ?	Clicks ?	Click-Through Rate ?
61,432	1.64	81	0.081%

The click-through rate gives you a way to look at performance. You might start out with a particular click-through rate, or you might try having multiple versions of an ad, with different images and text, and see which gives you a better click-through rate.

Targeting an Ad

Another central concept in ads is *targeting*. When you create an ad on Facebook, you can pick specific audiences, and the more targeted your ad is, the better it will perform generally speaking. In other words, the more your ad is focused on appealing to a particular audience, the better it will do.

Part of the targeting process is the ad itself. In a really simple example earlier, I included some phrases about the story I wrote, which might be of interest to people who like science fiction:

Free: Onym 1: Ant Onym
wattpad.com
Ant robots, neuromorphic chips, genetic artistry. Science Fiction, Professor Punk.

In the Targeting section of Facebook, I chose people who expressed an interest in science fiction in their Facebook profiles. I also chose a particular geographic area and language:

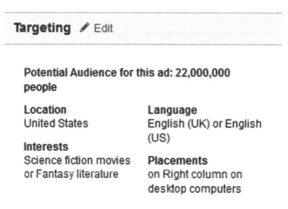

Targeting ✏ Edit

Potential Audience for this ad: 22,000,000 people

Location	Language
United States	English (UK) or English (US)
Interests	
Science fiction movies or Fantasy literature	**Placements** on Right column on desktop computers

Targeting is a way to match your advertisement to a particular audience.

Now for the fun stuff!

Creating a Campaign

To create a campaign, go to https://www.facebook.com/business/a/campaign-structure.

People also ask

What is a campaign on Facebook?   ⌃

Campaign: A **campaign** contains one or more ad sets and ads. You'll choose one
advertising objective for each of your **campaigns**. Ad set: Ad sets contain one or
more ads. You'll define your targeting, budget, schedule, bidding and placement at
the ad set level. Ad: The creative you use makes up an ad.

How to structure your Facebook Ads campaigns | Facebook for Business
https://www.facebook.com/business/a/campaign-structure

Search for: What is a campaign on Facebook?

What is an ad set in Facebook?   ⌄

Click on Create an Ad.

From time to time, Facebook will change its interface. At the time of this writing, the wizard allows you to choose from various kinds of ads. I suggest trying
the Page Likes or Clicks to Website options:

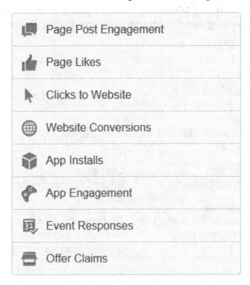

With the Clicks to Website ad, you provide the link to a web site you want to
send people to (such as the blog you created in Chapter 2).

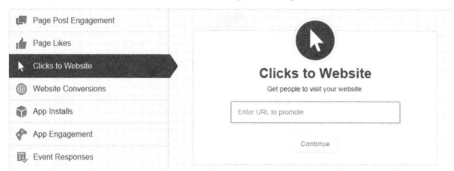

You simply enter the link and click Continue:

(Note: When you enter the link, it may flip you to another page even before you click Continue.)

On the next page, you'll have the opportunity to do some targeting, by choosing what audience you want the ad to display to:

As you make choices on the left side, the Audience Definition on the right side will be updated in real-time:

This is an interesting area to experiment with. To start, try entering language options.

Then try entering Interests. This is where you can really focus on a particular audience.

If you're keeping track of the numbers, you'll see that at the time of this writing, you start out with around 184,000,000 people you can send an ad to if you don't target them at all. These are the people who speak English in the United States who are also on Facebook.

As soon as you start entering interests, the numbers go down. People who have expressed a particular interest or liked a page that falls in a particular category, etc. There's a smaller potential pool, but this is a good thing. The ad is more targeted, so people who see it are more likely to click on it, at least in theory.

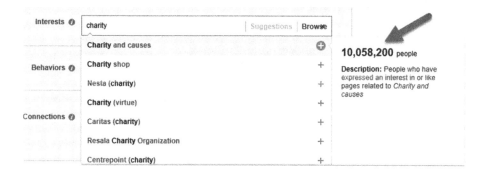

For now, I suggest skipping over Behaviors and Connections, but you might click on the little "*i*" circular icons in these sections to get more information.

Next you get into the Account and Campaign section.

Budgeting for Facebook ads is an art and a science. There are no universal rules, but a basic principle is that the more competition there is for a particular audience, the more the ads will cost.

One method for paying for ads is based on quantity, that is, the number of ads that are displayed. The other method is where you pay only for clicks that actually take place.

This is an area that periodically goes through changes, but the basic idea is you figure out a schedule—how much you're willing to pay—and then set it in motion.

And Now for Something Really, Really, Really Important

I think the most important point to keep in mind, especially when you're learning, is to know how to set a start and end date. That way, your credit card does not get continuously charged! Click on Set a Start and End Date to do this.

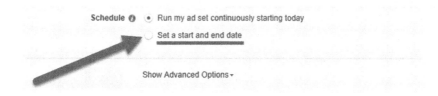

If you're just learning, you might want to try a week-long campaign.

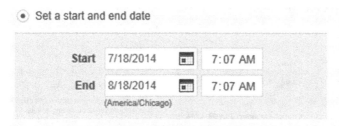

Instead of a suggested daily budget, you might click on the Per Day drop-down menu and choose a lifetime budget:

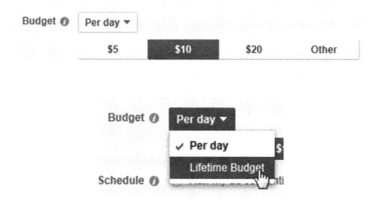

It will suggest a $50 budget, but you can click on Other:

You can type in something like $10-20:

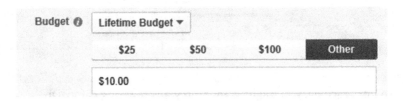

Be sure to set an end date:

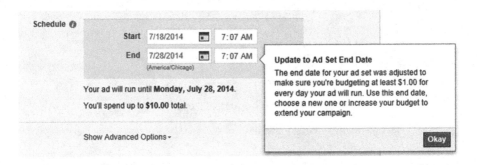

Click on Show Advanced Options to see what's there:

Show Advanced Options ▾

Facebook will automatically bid on ads for you. That means that it will look at your ad, the audience, and choose the best bid to get you clicks—that is, amount that will be paid for each click. On Facebook and Google, in the end, the advertising price is an auction process. You indicate a willingness to pay a certain amount, and depending on competition, the price will go up or down. You can always limit the maximum you will pay for a single click.

For example, you can click the Manually Set Your Maximum Bid for Clicks (CPC) option:

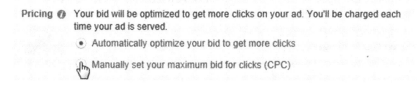

There is a little suggested range and you can accept the specific suggestion or change it.

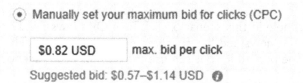

(You can also change this range after your campaign is running.)

Next, Facebook allows you to upload an image for the ad. In some cases, if you're creating an ad for a Facebook page, it will look on the page and automatically suggest the logo. With an external web site, it can also suggest images. This is a section that highlights the need for content (see Chapter 2—nudge nudge) and working with digital images. You might want to use images from clipart.com or work with them in Photoshop or a free program like GIMP, and then add text to the image. But to experiment, you can just use a digital picture.

Tip   If you know of an image that you want to use on the web, you can also right-click on it (in Windows) or Ctrl+click (on a Mac) and download it to your computer. Remember to respect any copyrights!

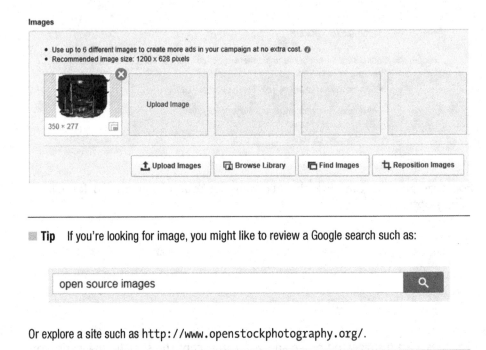

Tip   If you're looking for image, you might like to review a Google search such as:

> open source images   🔍

Or explore a site such as http://www.openstockphotography.org/.

Next, you can select the headline and text for an ad.

So in the Text and Links area, you can type in a headline and some text. Also, to keep things simple in this scenario, click Turn Off News Feed Ads. (An ad that appears in the newsfeed has to come from somewhere. That is, it will look as if someone posted it, so it can use the icon of your Facebook page.)

Text and Links

Connect Facebook Page

Choose a Facebook Page to represent your business in News Feed. Your ad will link to your site, but it will show as coming from your Facebook Page.

| Choose Facebook Page | ▼ | + |

or Turn Off News Feed Ads

Headline ❶   20

NPOEx

Text ❶   90

Enter compelling text that lets people know what you're promoting...

When you type a headline, it gives you a running count of how many letters you have left:

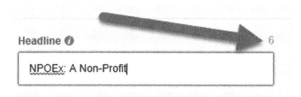

Headline ❶   6

NPOEx: A Non-Profit

The idea is to keep things simple. Offer a value proposition and a call to action of some kind:

Text ❶   6

Come and see a TEDx video about the idea of a non-profit stock exchange. Carpe diem!

Just look at what you're working on and consider what would get you to click on it.

On the right side of the screen, it's giving you some information about where the ad is going to appear.

This is the area where you can activate an ad to appear on a mobile device by clicking the Add link:

Finally, click Place Order:

(Note: If it is the first time you are making an ad, you'll need to enter billing information. Remember to make sure you're aware of how long the ad will be running.)

Monitoring the Campaign

The ad has to be reviewed and approved. You can see how things are going by visiting https://www.facebook.com/ads/manage.

You then click on the campaign name to drill down to the ad level:

After an ad starts running, this dashboard will show *reach*, which is how many people may have looked at it, and then it will show any clicks that occurred. At this point you will be looking at an *ad set*, which is a group of ads in the same campaign. You can create multiple ads with various images/copy/targeting and see how they perform against each other. To drill down further, click on the Ad Set name:

Then, to get a closer look at the ad, click on the ad's name:

	Status ?	Ad ?	Delivery ?	Results ?	Cost ?	Reach ?	Frequency ?	Clicks ?	Click-Through Rate ?
		Todd's story	Not Delivering Completed Website Clicks	101 Website Clicks	$0.34 Per Website Click	61,432	1.64	81	0.081%

Then you'll see the information and images you originally put into the ad. This is also where you can edit things if you want to:

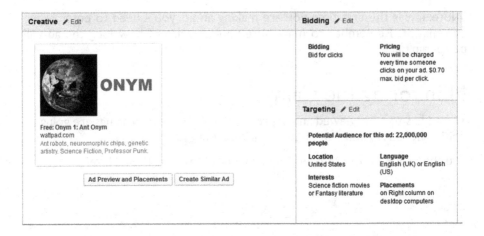

In this case, I wanted to generate some clicks. The performance I wanted to measure was how many clicks I got:

Clicks ?	Click-Through Rate ?
81	0.081%

I also wanted to see how much money was spent.

$34.93

In order to get about 80 people to come and see my short story, it cost me about 50 cents a click.

Learning More

There are a lot of resources out there for learning more about how ads work. Although I have a healthy skepticism for the ROI, it is definitely true that many businesses have benefitted financially from Facebook. As examples, see these success stories on Facebook:

- `https://www.facebook.com/business/success/maserati-usa`

- `https://www.facebook.com/business/success`

When you're wandering around an interface, try clicking on everything, especially the little question marks:

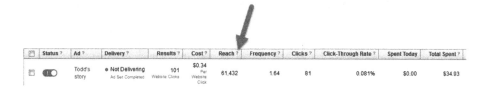

Status ?	Ad ?	Delivery ?	Results ?	Cost ?	Reach ?	Frequency ?	Clicks ?	Click-Through Rate ?	Spent Today	Total Spent ?
	Todd's story	● Not Delivering Ad Set Completed Website Clicks	101	$0.34 Per Website Click	61,432	1.64	81	0.081%	$0.00	$34.93

To learn about Reach, click on the question mark next to it, for example:

61,432

Additional areas for information include https://www.facebook.com/business/products/ads. The main help section also includes a program called "Start to Success" that you might be interested in. See https://www.facebook.com/help/458369380926902.

Free Facebook Ads

You can also explore promotions that companies offer that give you free Facebook advertising credit. For example, try entering the following in Google:

Conclusion

Congratulations on making it through the chapter. The best advice I can give on learning about Facebook ads is to explore them, try them out, and spend some time in the help sections on Facebook. I also suggest keeping things simple. It can be a lot of fun to see the clicks start coming through the first time you try, which can naturally lead to learning more about the features and options that help make your ads perform better.

If you're an intern or student, try to find a web site and set the goal of seeing how many clicks you can get. Then try again to see if you can come up with more compelling images, or text, or more focused targeting. See what your original click-through rate was and see if you can improve it.

You may want to start asking the question, "what's my ROI?". If you're interested in selling something, you'll want to start looking into a topic called *conversion*, which is tracing an ad down to the point of actually selling something.

Twitter

This chapter looks at Twitter as a social media marketing channel, discusses its pros and cons, and walks through the process of creating a Twitter account.

Twitter: To Do or Not to Do?

Although I believe it's wise to see the social media marketing landscape as prone to shifting, Twitter is considered to be a "mainstay" channel. It's one of the networks that companies often post to. As with Facebook, there's no right or wrong way to use it, or not use it. That said, the best questions to ask is what audience you want to address and is that audience on Twitter?

As with Facebook, people are generally less likely to have "purchase intent" when on Twitter (as compared to searching on Google), but this could change. Just like Facebook, Twitter has explored advertising. A fair amount of people spend time on Twitter, and they're generally a younger demographic and probably more prone to being interested in exchanging news and information.

It's also fair to say that for the segment of the population that uses Twitter, it's increasingly common to "tweet" (post something) about issues that a person is having with a product or service. When discussing social media marketing, you invariably end up having to consider the conversation as "two-way". Even if you don't advertise on Twitter, you may want to become familiar with how people are tweeting about your company or organization. The bigger the company is, the more likely people will expect that if they tweet about an issue they're having, the company will join in on the conversation from a customer service perspective.

© Todd Kelsey 2017
T. Kelsey, *Introduction to Social Media Marketing*, DOI 10.1007/978-1-4842-2854-8_5

My general recommendation is to become familiar with Twitter, to look at some of the techniques for developing and maintaining a presence on Twitter, and to think critically about its ROI, as with other channels.

If you're in the position of pitching or developing a baseline social media presence for a company or organization (or person) that knows they want to be on social media, it might be helpful to think about Twitter as one of several channels that you can post to.

■ **Note**   Start with a plan of posting regularly to a blog or help a business owner or organization post regular blog articles, and then work to post those blog articles to Facebook and Twitter, in order to maintain a presence on these channels.

It's a bit hard to make generalizations on the effectiveness of a particular social media channel, but I do think that Twitter is at least worth trying. Be sure to at least monitor it and become familiar with it.

A High-Level Overview of Twitter

To understand how Twitter works, you can look at a Twitter page without being "on" Twitter. For example, take a look at `www.twitter.com/reuters`.

This is an example of a very common use of Twitter—to exchange the latest news about whatever (in this case, it's Reuters, an official news organization):

Twitter posts are limited to 140 characters, so the general format is some kind of very concise comment, often with a link to an article or picture, and then some conventions that are somewhat unique to Twitter.

For example, in the previous graphic in the second post, there is @reuters – this is like a "ping". If I wanted to mention Reuters in my own tweet, if I include @reuters, Reuters can scan Twitter and see that the post mentions their username. The @ sign is basically a way to notify another Twitter user about something connected to them.

Another common thing on Twitter, which you'll sometimes see on television news or in media discussion, is a *hash tag*, such as #computers or #apple. It's basically another kind of tag. If you include this kind of text in a Twitter post, you are associating the post with any other post that is using the tag. So hash tags will commonly become popular around events or popular topics, such as #worldcup, or around celebrity names, etc.

In addition to news, Twitter is often used by companies and brands. For example, Oreo has an active Twitter page at www.twitter.com/oreo.

Are they posting about real "news"? Not particularly; just seeking to make interesting posts.

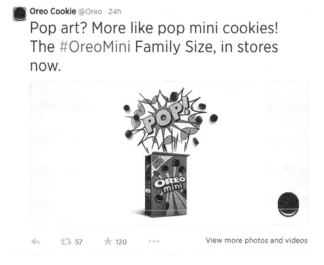

As much as I believe that it's important to think about return on investment in social media, it's fair to say that a good portion of what goes on in social media basically amounts to show and tell, where effective posts can simply be entertaining. They can catch someone's attention because they're bored, not necessarily because they're looking to purchase something.

It's also fair to say that Twitter is used extensively for entertainment purposes. Here's a list of some top Twitter accounts at the time of this writing:

Twitter users		Followers	Following	Tweets
1	KATY PERRY @katyperry	54,597,665	149	5,807
2	Justin Bieber @justinbieber	53,031,491	133,102	27,306
3	Barack Obama @BarackObama	44,200,798	649,397	12,145
4	YouTube @YouTube	43,785,293	712	10,797
5	Taylor Swift @taylorswift13	42,000,315	130	2,265
6	Lady Gaga @ladygaga	41,648,832	134,306	4,924

Another area to be familiar with is customer service. Many people tweet about products, and it is a best practice for a company to keep an eye on social media. An upset customer is more likely to start a firestorm on social media, and it can quickly get out of control if you don't address it.

Some are useful because they offer deals and customer service. Others are just plain entertaining.

Here's a sample post that someone made. Notice how they referenced @BofA_Community—it's the "ping" I was talking about. If the Bank of America is paying attention, they could respond.

Bank of America did respond, because they have a social media team who is out there keeping an eye on things. Keep in mind that these days there are people out there who won't bother e-mailing or calling a business, but they will tweet. Tweeting does not necessarily mean going on a web site and entering a post—many people tweet right from the text message function on their phone. This is part of the reason that the Twitter app spread so fast—because it is fairly simple and you don't even need a computer to use it.

Creating a Twitter Account

To start exploring Twitter, go to `twitter.com` and enter your name, e-mail address, and a password.

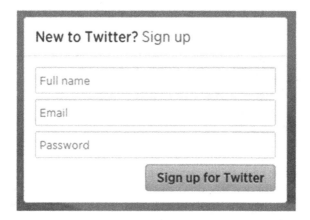

Next it will flip you to a screen where you can choose a username. This will become `twitter.com/yourusername`.

Choose your username

Don't worry, you can change it later.

Keep trying usernames until you find one that's available and you like:

Then, click Create My Account:

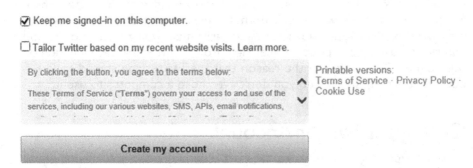

Just like Facebook does, Twitter regularly changes or "improves" the way things work. At the time of this writing, the sign-up process seemed to force you through a tutorial, so you can just keep on clicking on the Next button:

It helps you to see how things work—in this case it is showing you how to "follow" people. In theory, you don't need to follow anyone, but when you find accounts to follow, they are notified, and often will follow you as well. One way of building a Twitter following is to find other accounts to follow. Whether people actually read each other's posts is another question that depends on how relevant and interesting they are to the person. But that's the basic part of it.

You can just click on a few people it suggests:

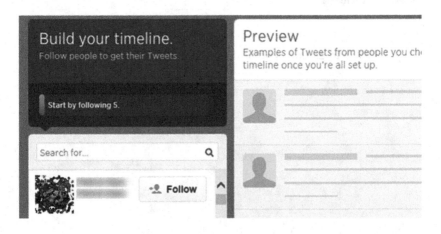

At some point, you'll get a confirmation e-mail:

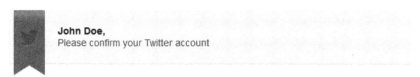

You just need to click on the Confirm Your Account Now button:

If you get annoyed with the step by step wizard, you can always go directly to twitter.com. It will probably look something like this, with some tweets on the right side from people you are following, and a "promoted" tweet (see the arrow). Promoted tweets are a form of advertising on Twitter that you can also explore.

To get to your own Twitter page, click on the profile icon/name:

Posting to Twitter

To get to your page, you can sign in and click on your profile icon, or you can go to `twitter.com/yourusername`, such as:

`https://twitter.com/mediasasquatch`

If you haven't yet, you can add a photo or follow the introductory suggestions:

In general, to tweet, you just go to the upper-right corner of the screen and click on the Tweet icon:

For example (nudge nudge), if you created a blog as suggested in Chapter 2, you can tweet your blog's address:

Twitter will give you a running count of how many characters you have left in the tweet. If you run over, Twitter will split it into an extended post or into multiple posts, and you'll generally want to keep posts under 140 characters.

After you enter the text, click Tweet and voila! You have posted to Twitter:

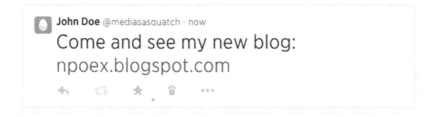

Then, in theory, as with newsfeeds in Facebook, the tweet should appear in someone else's Twitter feed. I believe, until/unless things change, if a person were to log on to Twitter and read every tweet from someone they were following, they would see yours. Unlike Facebook, you don't need to "promote" a tweet in order for someone to see it. But don't be surprised if Twitter goes from theoretically unrestricted access to tweets to "monetizing" them in some way.

ROI and Strategy of Twitter

An entire industry has grown up around marketing on Twitter, including leveraging advertisements, promoted tweets, etc. If you're just learning about what's possible, or you want to know what kinds of services could be offered to a business or organization, I recommend taking a look at these sites:

- `https://business.twitter.com/marketing-twitter`
- `https://business.twitter.com`

In terms of return on investment, at least in theory, if your business is selling a product, you can include promotions, coupons, and offers, just like you can in an offer on Facebook. Part of the science of tracking ROI in that case is including links that allow you to calculate the conversion rate—"Of the number of people who came to our site from Twitter based on that link, how many actually purchased something"?

Learning More

Twitter has some good learning material at `https://support.twitter.com`.

I also recommend taking a look on Google at topics such as "Twitter marketing" and "ROI on Twitter", something like "Twitter best practices", or "leveraging Twitter for customer service".

You can also explore these links:

- `http://www.businessinsider.com/best-brands-to-follow-on-twitter-2013-9`
- `https://blog.twitter.com/2013/guardian-says-twitter-surpassing-other-social-media-for-breaking-news-traffic`
- `http://twittercounter.com/pages/100`

Conclusion

Congratulations on learning about Twitter! As with some of the other social networks, the best way to learn is to try it a bit and to look at how other people are using it. One of the common things that happens with Twitter (and Facebook for that matter) is that an individual, business, or organization starts out with a "gung-ho" approach, spending lots of money and time. The regularity of posting will then drop off, and their social media presence will languish. This is why, in spite of how important it is to become familiar with Twitter, it's helpful to consider that it still comes down to content. Ask, what are you going to post? A sustainable, ongoing content strategy should be a central part of any social media effort.

Yes, establish a social media presence for yourself or for your business, but not without a realistic, sustainable plan for regular posts—whether those posts are offers, news, promotions, or just share information about the organization. I highly suggest starting out with modest goals, such as a monthly or weekly post, and when you've regularly posted for a while, consider posting more. Over the long term, it's better to post regularly, at higher quality, and with more attention. I guess I'm a believer in blog posts for that reason.

LinkedIn

This chapter takes a brief look at LinkedIn and talks about how it can be incorporated into a social media marketing strategy, including looking at LinkedIn pages, which can be created for a company or for an individual. We also touch on LinkedIn groups and cover LinkedIn's increasing options for advertising.

LinkedIn: To Do or Not to Do?

In general, establishing a presence on LinkedIn is probably most useful to a B2B (business-to-business) company, where your customers are other businesses or organizations, including people who are potentially more likely to be spending time on LinkedIn. As a social media marketer, the priority might be for you to strengthen your own profile on LinkedIn. For example, you can create a LinkedIn group, and it can be a way for people to discuss a topic. In order to build a network for finding work or clients, you might want to search for and join LinkedIn groups and work on adding examples of projects you to do expand your LinkedIn portfolio.

In other words, the suggested prerequisite for considering LinkedIn as a social media channel is to first get in the habit of strengthening your own presence. Not only will it help you with your career, but if you do end up getting involved in establishing a presence on LinkedIn for a client, people will most likely look at your profile, especially if you create or interact in LinkedIn groups.

My general, strong recommendation is to explore LinkedIn. Maybe set aside a half hour at least once a month to look at your profile. Build it up a bit and get in the habit of getting involved on LinkedIn. Work toward thinking of LinkedIn when you create a new project or work on one and posting the results to your profile.

© Todd Kelsey 2017
T. Kelsey, *Introduction to Social Media Marketing*, DOI 10.1007/978-1-4842-2854-8_6

In terms of general social media marketing, I also recommend getting at least somewhat familiar with the options LinkedIn has for B2B social media, including how a company/organization page can be created, as well as looking at some of the advertising options LinkedIn has.

It's not strictly B2B (business to business), as there are plenty of companies and organizations that have at least some LinkedIn page presence, even if it's passive. Increasingly, even in general and "B2C" (business to consumer) situations, LinkedIn is used for recruiting, posting jobs, etc. It's a good idea to become familiar with it. Even if you're in a large organization, don't "assume" anyone is managing LinkedIn. As a social media marketer, you might be able to explore how it can be leveraged.

As with other channels, I think it's important to consider ROI. With LinkedIn in particular, it has become a default hub for businesses, non-profits, and just about any job seeker, so at a minimum, your organization should have a page and someone making sure the information is accurate. They should be posting updates once in a while, or at least posting jobs, opportunities, projects, etc.

The importance of this increases tenfold if your employer or client is a B2B entity, where people may look you up on LinkedIn as a company. It's also true that when people search on Google, it might turn up the company or organization page. It's a good idea to keep an eye on it and potentially leverage it.

For example, here's an example of a familiar company and their LinkedIn page:

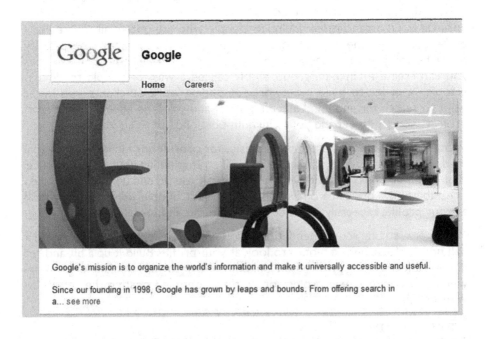

Just like with other social networks, you can make posts and provide updates:

Recent Updates

Google Today more than 190 million people actively use Google Drive at home, school and work. Drive keeps all your work safe, and makes it available everywhere and easy to share. And today we launched Drive for Work, new tool for businesses that includes ... more

Google Drive for Work

goo.gl · Google Drive for Work comes with unlimited storage, 5TB file uploads, advanced audit, eDiscovery and more. All your work, safe, available everywhere and easy to share.

Like (1,355) · Comment (115) · Share · 27 days ago

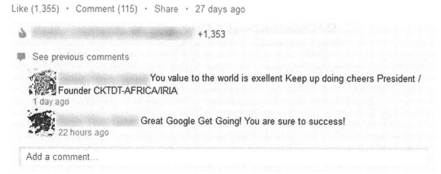

+1,353

See previous comments

You value to the world is exellent Keep up doing cheers President / Founder CKTDT-AFRICA/IRIA
1 day ago

Great Google Get Going! You are sure to success!
22 hours ago

Add a comment...

Understanding LinkedIn

To understand LinkedIn, the best thing to do is use it regularly. If for some reason you don't have a LinkedIn account, I highly recommend you get one. If you have one but don't have many "connections," I invite you to go on to a place like Vista Print, go for the free business cards (where you pay for "shipping only"), and consider making a business card with basic contact information, including your LinkedIn Profile link.

The reason I suggest a simple business card (even if you have an "official" one) is to just have some around, so that when you think of it, you can get in the habit of inviting people to connect with you on LinkedIn. You'll also want to get in the habit of inviting people by e-mail, such as colleagues at work, or clients, or alumni if you went to college. Don't overlook relatives, neighbors, and people you see each week.

The general value of taking a bit of time and expanding your network on LinkedIn is that it can be a helpful, very valuable tool for connecting to jobs and new clients. For example, you might want to contact someone to ask about a particular project, job, etc. Although you might not know them personally, someone you know does, as you can see from LinkedIn. LinkedIn basically helps you connect to people, which for many people, has a more powerful, sustainable impact than Facebook in establishing networks. But don't be shy if you've never tried it!

Note Get on LinkedIn, get a Gmail address if you don't have one (mail.google.com)—because Gmail rocks—and then learn how to use Google Calendar. Set a monthly reminder to spend half an hour on LinkedIn. That's my suggested way to get started.

To get on LinkedIn, go to linkedin.com and sign up!

Get started – it's free.
Registration takes less than 2 minutes.

First name	Last name

Email address

Password (6 or more characters)

By clicking Join Now, you agree to LinkedIn's **User Agreement**, **Privacy Policy** and **Cookie Policy**.

Join now

Creating a LinkedIn Page

Actually creating a LinkedIn page requires you to jump through a few hoops. It's not like creating a random web site—it really does need to be official.

Having said that, if you're learning social media marketing, I recommend considering creating a web site for yourself, with an "official" web site name, such as blahsocial.com/net/, and getting an "official" e-mail address (you@blahsocial.com). In part, this will be helpful as an exercise in learning how to set up an Internet presence, period.

In order to do this, I recommend exploring the least expensive web site plans at places like godaddy.com and 1and1.com, and making notes, even calling up and talking to a live person, about the cheapest way to make a web site with your own name.

If you want to try the Google route and have the cheapest monthly cost, you can sometimes just register the web site name (such as blahsocial.com), which includes e-mail capability, and not use godaddy.com or 1and1.com for hosting the web site. Instead, you can point the web site name at a free service, such as Google Sites. I also recommend looking at strikingly.com, wix, weebly, and shopify. Just try them out. After you've tried all the free ones, choose one for your "official" site. It's good to become familiar with them all in case a client wants help in establishing an Internet presence. If you're employed by a larger company, it can still come in handy to know how to create a microsite based on a particular campaign.

You can start by glancing through this section to understand in general what you need to do to create a LinkedIn page. Once you have your official site and an official e-mail, come back and try making a LinkedIn page.

To get started making a LinkedIn page, visit https://www.linkedin.com/company/add/show.

Add a Company

Company Pages offer public information about each company on LinkedIn. To add a Company Page, please enter the company name and your email address at this company. Only current employees are eligible to create a Company Page.

Company name:

Your email address at company:

☐ I verify that I am the official representative of this company and have the right to act on behalf of my company in the creation of this page.

Continue   or Cancel

If you try to wing it, LinkedIn will politely refuse:

❌ Sorry, gmail.com is not a valid company domain. For further questions, please contact us using the Customer Service link at the bottom of this page.

Basically, LinkedIn is looking for an "official" web site address, based on a web site name and an official e-mail address.

> **Tip**   Even if you have an official e-mail address, you can still check it "through" Gmail. I believe Gmail has the best, easiest, most flexible framework for searching/managing e-mails, as well as the best spam filtering. You create a Gmail account and then add e-mails in the Settings area. Google also enables you to have your own private Gmail interface for an official web site name. In other words, you get all the Google tools, but under `yourname.com`. I worked at a startup that used this approach, and it was very helpful. Inquiring minds should look into Google Apps. It might very well be the kind of thing that a new client, a non-profit, or any sized business would be interested in for that matter. For example, many colleges are using Google Apps increasingly, so that students can use the Gmail interface for the `school.edu` address. It can often be cheaper and less hassle for IT to manage a Google Apps account, than have to manage all the moving pieces of other e-mail systems.

Eventually, you'll end up entering something like this:

Add a Company

Company Pages offer public information about each company on LinkedIn. To add a Company Page, please enter the company name and your email address at this company. Only current employees are eligible to create a Company Page.

Company name:

RGB Green

Your email address at company:

todd@rgbgreen.org

☑ I verify that I am the official representative of this company and have the right to act on behalf of my company in the creation of this page.

[Continue] or Cancel

Then you get a confirmation e-mail:

 Thank you. A confirmation email has been sent to (info@cftw.com).

Now check your inbox...

To continue creating the Social Sasquatch profile, please do the following:

1. Check your inbox for **info@cftw.com**
2. Follow the instructions in the email
3. Complete your company profile information

You'll need to confirm it.

Hi Todd,

<u>Click here</u> to confirm your email address and finish creating the Sasquatch Media company profile.

If the above link does not work, you can paste the following address into your browser:

You will be asked to sign into your account to confirm this email address. Be sure to sign in with your current primary email address.

There are standard pieces to starting a company page, such as having a logo and basic information. (Read Chapter 2 on content and learn how to work with digital images or go to 99designs.com and get a logo the crowd sourced way.)

Posting to the Page

LinkedIn pages can be pretty straightforward; you just go on and share an update.

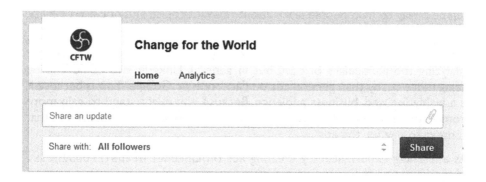

You can think of the LinkedIn page as a directory listing, in a way.

Specialties
free learning material in different languages, multilingual open source CMS, intellectual property management and development, corporate social responsibility

Website	**Industry**	**Type**
http://www.cftw.com	Information Technology and Services	Public Company

Company Size	**Founded**
1-10 employees	2012

Something Really Interesting and Important

One of the interesting things to me is that a tool like Hootsuite, which we'll look at in a later chapter, allows you to post simultaneously to multiple networks. I worked at a B2B startup, and they wanted to actively engage across multiple social media channels. I would make a blog post which was the central, official place for new content. Then, using Hootsuite, I automatically posted to Facebook, Twitter, and LinkedIn. In the case of LinkedIn, it wasn't that there had to be a lot of work put into it, but when the right pieces were put together, it was relatively easy to post content to it.

ROI Strategies for LinkedIn

So how do you calculate ROI on LinkedIn? Technically, if you built up followers, you might post offers/deals once in a while, and you could theoretically track what resulted. Another fair way to think about LinkedIn is as an outlet for PR, or just another way to build credibility with relevant content. This is especially true in a B2B setting, where developing content and sharing it on social media can be a way to help people learn about relevant topics and boost credibility.

LinkedIn also has advertising, and in some cases, especially in a B2B setting, the overall cost of a product or service might be higher. I'm not tracking or analyzing the particulars of cost, but in general, LinkedIn ads could end up being more costly than general ads on Facebook or Google. This is because it is a highly focused, business audience. Because you might stand to gain more from a successful sale in a B2B setting, there's probably more appetite in that context to pay for the ads.

As a learning experience, try creating ads on LinkedIn: `https://business.linkedin.com/`

Then explore the marketing side: `https://business.linkedin.com/marketing-solutions`

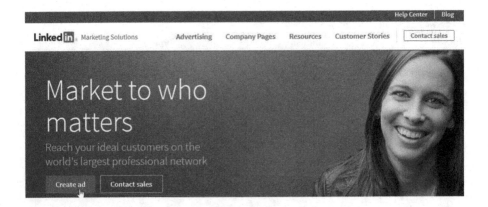

Personal LinkedIn Development

I also recommend taking a look at these links to get more information about how to build your own LinkedIn presence:

- `http://jobsearch.about.com/od/linkedin/ss/linkedin-profile-tips.htm`

- `http://www.inc.com/janine-popick/5-easy-steps-to-get-started-on-linkedin.html`

- `http://www.powerformula.net/free-resources-for-learning-linkedin/`

- LinkedIn Learning Webinars: `https://help.linkedin.com/app/answers/detail/a_id/530`

Learning More

Unfortunately, LinkedIn's help section leaves a bit to be desired. It makes the mistaken assumption that you know what questions to ask, rather than having an easy to find Getting Started section. However, this link is worth exploring in relation to specific pages:

`https://www.facebook.com/help/364458366957655/`

This link also has some general information that's worth reviewing:

`http://business.linkedin.com/marketing-solutions/company-pages/get-started.html`

Conclusion

Thanks for reading! LinkedIn is an important tool for social media marketers, including for building your own presence and understanding how to set a company or organization up with its own page. If you intend to do freelance, this can also be a service. In other words, something that may be overlooked, or something you could do for a local business or organization to get experience.

You might also want to search for other social media discussion groups on LinkedIn. It can be a nice badge on your profile and a way to learn about new things.

Hootsuite

In this chapter, we're going to look at Hootsuite, a tool that helps you manage social media by providing assistance on developing and posting content to multiple social networks. It's part of the "skill toolbox" that I recommend exploring. If you haven't created a blog, Facebook page, Twitter account, and LinkedIn page yet, I recommend reviewing the first chapters in this book and creating those accounts. You can get something out of this chapter simply by reviewing it, but I recommend setting up those accounts and then "connecting" them to Hootsuite. The LinkedIn page isn't strictly necessary, but I recommend setting the goal of having 2-3 social media accounts to post to.

I also recommend considering your blog as "home base" for social media, the central place where you develop and post content initially, which you then post to various social media properties. Part of the reason for this is because with a blog, it's literally "your" social network. Unless Google or WordPress goes down in flames, your blog will be around for as long as you want it, and *you* control it. You could argue the same for Facebook and Twitter and LinkedIn—at least in terms of the likelihood of them continuing—but the primary purpose of a blog is content, and you have much more control over things.

At any rate, if it all sounds too abstract, the best way to try things out is not just to develop a social media presence, but to develop one and see what happens when you post content on an ongoing basis. Then you'll start to see the satisfaction and excitement when people comment on your blog post, like your post on Facebook, get more followers on Twitter, and so on. You may also encounter what thousands of people have encountered—a burst of initial enthusiasm, audacious goals, and then getting busy (or overwhelmed) so that the amount and regularity of content trickles off. (I'm guilty of this myself! Don't ask me how many blogs or web sites I have!)

© Todd Kelsey 2017

T. Kelsey, *Introduction to Social Media Marketing*, DOI 10.1007/978-1-4842-2854-8_7

This is why over time, I've tended to gravitate toward free tools. With paid web site accounts, it might look nicer, but in some cases, sustainability is more important than looking nice. It's better to have a free site you don't have to pay for, than one you have to struggle with and decide whether to pay for on an ongoing basis. I'm primarily speaking of personal sites, because when you get into business or non-profit sites, it probably makes sense to have the most professional site you can. But I always recommend considering what is sustainable.

Since social media is, in my opinion, primarily about content—show and tell—I think it's important to think about what is sustainable in terms of time spent, and return on investment, from the beginning. At the very core, anything that makes it easier to develop and post content is a good thing. This is why Google's tools are so good, because you spend less time and hassle fiddling around and can focus on the content.

Hootsuite fits right in the middle of that. When you develop the content, Hootsuite helps make it hassle free and easy to post content to multiple networks, including scheduling things ahead of time.

Understanding Hootsuite

To get a sense of the kinds of things you can do in Hootsuite, I recommend going to www.hootsuite.com and glancing at some of the menus at the top—Products/Solutions/Plans/Services.

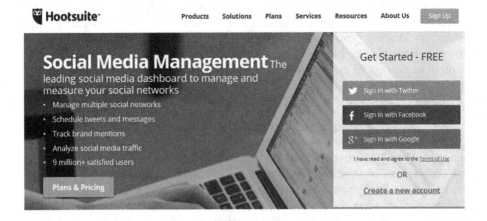

It's not critical to understand all of the options. (Remember, in an earlier chapter, the discussion about choosing some "doable" things—focusing on learning some basics—so that you don't get overwhelmed.)

The Plans and Pricing link is worth looking at and it's nice to know there's a free version. In general the tool revolves around managing the creation and posting of content. It can also be used for *social listening*, which helps you get a sense of what people are saying about you on your social media sites.

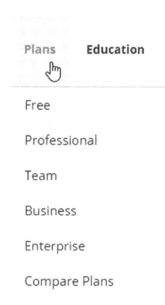

Hootsuite on YouTube

Hootsuite on YouTube can also be a nice place to start, to watch a few videos on social media in general, and on Hootsuite in particular:

https://www.youtube.com/user/hootsuite

Hootsuite Help

The Hootsuite Help section is also worth exploring, including the Getting Started link at https://hootsuite.com/help.

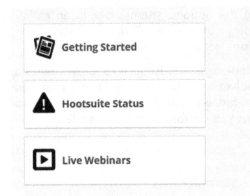

It gets down to the basics and can help you understand some of the things you can try. Later in the chapter, we'll go through creating an account and posting a message. Installing apps and "going mobile" might be something you'll want to wait and try after you have mastered the fundamentals.

Hootsuite's Quick Start Guide summarizes 6 important steps that will help you get started with your Hootsuite dashboard.

Step 1: Sign Up for Hootsuite

Step 2: Add Social Networks

Step 3: Add Tabs and Streams

Step 4: Compose and Send Messages

Step 5: Install Apps

Step 6: Download Hootsuite Mobile

In the help section, there's also a link called Live Webinars:

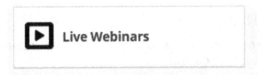

You can also get to the Live Webinars page using this link: http://media.hootsuite.com/webinars/.

If you scroll down, there are also on-demand (recorded) webinars:

They have a wide selection of webinars that is always evolving, so check it out.

Social Media Trends You Need To Know For 2016	Introduction to Hootsuite: Part 3	Introduction to Hootsuite: Part 2	Introduction to Hootsuite: Part 1
	Social listening to get you started	Taking advantage of Hootsuite's scheduling tools	Easily build your Hootsuite dashboard
GENERAL	MONITORING		GENERAL

Creating and Configuring a Hootsuite Account

Let's get started creating a Hootsuite account!

Note   If you're already signed in to any of the social networks (such as Facebook, Twitter, or LinkedIn), the screens may look different, but the setup process is fairly straightforward. Just like all the other social networks, the screens seem to change periodically. Never fear, just think of it as an adventure!

To get going, go to hootsuite.com and click Sign Up:

I suggest just starting with the free account for now, even though there's a free trial of the more advanced account.

Just click Get Started Now:

Then enter your e-mail address, name, a password, and your location, and then click Create Account:

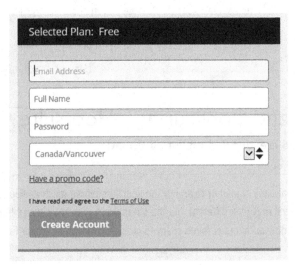

Next Hootsuite will walk you through the process of adding social networks. This means establishing a connection to social media accounts you have, so that you can use Hootsuite to post to them. If you don't have any "official" accounts yet, you can always add them later, but to get started, I suggest adding three personal accounts. For example, I added my Twitter, Facebook, and LinkedIn accounts.

To add social networks, just click on them:

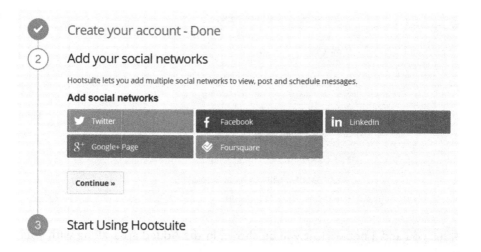

Then you'll need to authorize Hootsuite to connect to the account. If you're already signed in to any of the networks, the authorization may be different, but you'll probably need to sign in. You fill out your info and click Authorize App. It will also make your life easier if you click Remember Me, but don't do this if you're registering on a public computer:

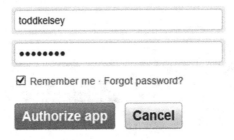

Authorize Hootsuite to use your account?

This application will be able to:

- Read Tweets from your timeline.
- See who you follow, and follow new people.
- Update your profile.
- Post Tweets for you.
- Access your direct messages.

> toddkelsey

> ●●●●●●●●

☑ Remember me · Forgot password?

Authorize app **Cancel**

This application will not be able to:

- See your Twitter password.

After you add a network, it will be shown in the Added area, along with your profile icon:

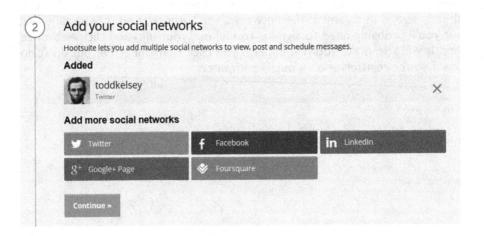

Next, try Facebook:

f Facebook

Log in to use your Facebook account with HootSuite.

Email or Phone:

Password: ••••••••

☐ Keep me logged in

Can't log in?

Click Okay when you're ready:

HootSuite would like to manage your ads and access your Facebook Pages' messages.

Skip   Okay

With Facebook, it will give you a list of items that Hootsuite can post to, starting with your personal profile, and then any Facebook pages you created.

At this point, if you haven't created a Facebook page (see Chapter 3), you might want to do that. However, you can always add one later. You can also add your personal profile and see what it's like to schedule a post to that.

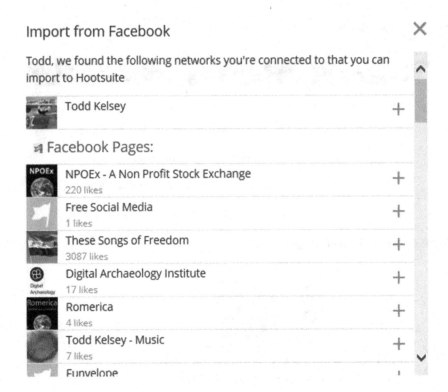

I recommend starting with a single item from Facebook, either your personal profile or a Facebook page. (Hootsuite limits you to posting to a total of three "items" across three social networks in the free version.)

Click the + sign next to an item you want to post to.

Then click Finished Importing:

Finished Importing

Next, try adding LinkedIn. Even if you haven't created an "official" LinkedIn page, you can always post to your LinkedIn profile. Regularly sharing blog posts and posting them to Facebook, Twitter, and LinkedIn can be a good way to get things out there. It can be as simple as an account of your adventures trying different things and sharing projects when you create them.

Adding social networks doesn't mean you always post to them—you have total control of what you post to and when. It's just giving you options. Hootsuite is a big timesaver.

Try clicking on LinkedIn if you like:

(Note: The free version of Hootsuite allows three social networks.)

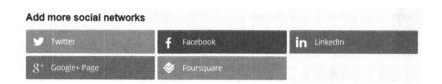

Then you'll need to enter your login info for LinkedIn and click Allow Access:

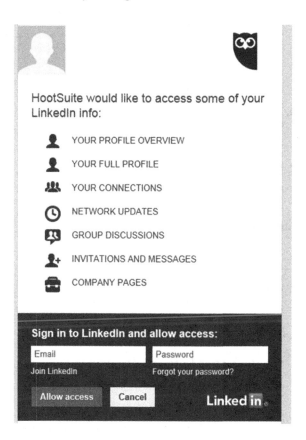

You should end up with a screen that looks something like this, with a list of networks/pages you added.

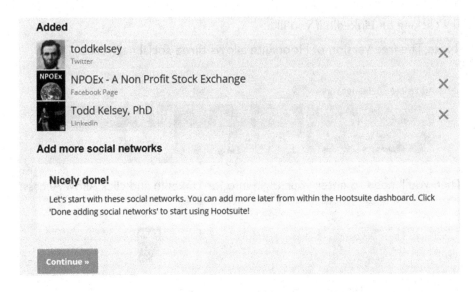

Click Continue:

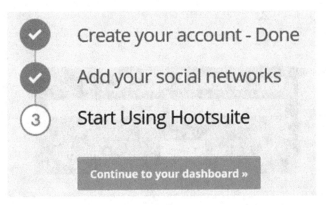

Then click Continue to Your Dashboard:

Configuring Hootsuite

In an attempt to be helpful, and as part of the setup process, Hootsuite will also take you through a series of steps to get started. I don't think there's any harm in going through the wizard, but you can skip it at any time (especially if you're just starting out and your eyes start to glaze over):

If you click Get Started, it will suggest adding *streams*, which are basically Hootsuite's "social listening" capability. If you want to know more about that, see the "Learning More" section at the end of the chapter and check out the related social listening video.

Basically Hootsuite will set up a feed of people who are mentioning you.

The way the wizard is set up is that you add a stream of a type of social media post—not ones you make in this case, but ones other people make. The type of item will depend on what network you added, but if you added Twitter, it will look something like the following screen (you can always click Skip Tour).

To add a stream, click on something like @Mentions (this is when someone on Twitter uses @yourusername, which is one way of communicating on Twitter—see the "Learning More" section of Chapter 5 for more information about Twitter).

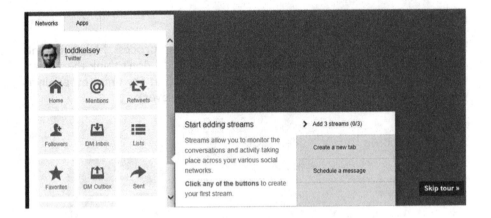

It will turn green. Then click on Followers and then on Scheduled:

This allows you to see what's happening on your Twitter account at a glance. The Scheduled stream allows you to see what posts you scheduled.

Next, you can add tabs:

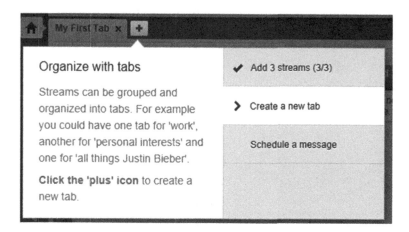

At this point, I just suggest clicking Skip Tour. We now go into Hootsuite as it normally appears and learn more about the interface:

Depending on what you've added, the "stream" screen will look something like this:

Getting Around Hootsuite

The main way you get around Hootsuite is through the toolbar on the left side:

When you roll your mouse over it, it will expand.

As with any social media tool, there are a variety of options with Hootsuite, but I think the best one to get started with is *simple posting*, which is based on the toolbar at the top:

Posting with Hootsuite

To make a post in Hootsuite, go to the toolbar at the top of the screen and click on the little drop-down triangle to choose where you are going to post:

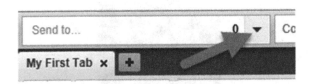

The social networks you added will be listed there. Just click to select them:

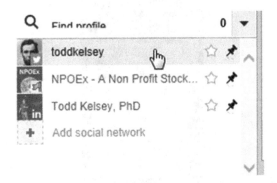

For example, this view shows that I am going to post to Twitter and the NPOEx Facebook page:

To get a sense of how helpful this is, you may very well want to enter into the world of social media marketing by manually posting things. Go ahead and sign in to Facebook, and then in separate tabs, sign in to Twitter and LinkedIn. Try manually posting and signing in to all those social networks. Pretty soon you'll see how nice it can be to have a central place to easily post to all of them.

When you are using the Quick Compose area (roll over it at the top of the screen), and you choose the social networks as we've shown, there's an interesting indicator at the bottom. It shows how many characters you can post to each network (the length of the message).

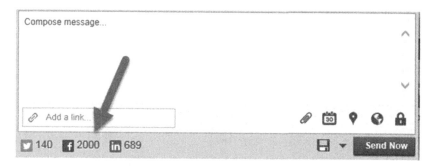

Try typing a test message by clicking in the Compose Message area and clicking the Send Now button:

You should get a Message Posted button:

If you post a message that's longer than Twitter can handle (140 characters), only the first 140 characters will appear on Twitter, along with a link to the entire message.

After you post your message, you can check and see how it will appear:

`https://twitter.com/toddkelsey`

You can also go on Facebook:

Voila!

I had trouble with the post appearing on LinkedIn. In theory, it should appear when you click on the Home button:

But posting to LinkedIn didn't work for some reason, until I tried clicking only on LinkedIn:

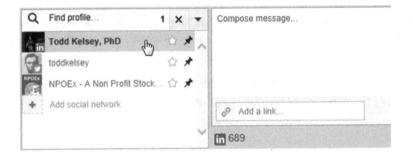

Then it appeared just fine:

The Joy of Scheduling a Post

This is another area where you could try using social media for a while. Try manually posting at different times of the day or week. You could do it manually to multiple networks. When you're working with social media marketing, you want to keep a regular stream of content going, whether it is content you develop yourself or an interesting article you find.

Tip For some thoughts on how often to post, see http://www.nimble.com/blog/posting-and-analyzing-on-facebook/ and http://blog.bufferapp.com/social-media-frequency-guide (you can also try searching on Google for how often to post on social media).

If you try to do it manually, it can be easy to forget, and it becomes uneven, and sometimes even trails off. This is a fundamental issue with many social media marketing efforts. There is initial enthusiasm, maybe posting a bit too much, and then it drops off, because of being busy with other things. I suggest using Hootsuite to accomplish the opposite result. You slowly starting out with some regular posts—maybe once a month or once a week.

When you get a "pipeline" of content going, it can be really nice to schedule things in advance. That way, you can work on developing content and get a week's worth of work done more efficiently.

To schedule posts in Hootsuite, click on Compose Message at the top, type in something like "hello again" as a test, and then click the little calendar icon:

Then you can choose a day/time for your post. Then click the Schedule button:

You should get a confirmation message:

Message scheduled

You can always roll your mouse over the toolbar on the left and click Publisher to check on your scheduled posts:

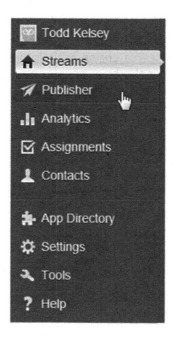

The posts will appear in a list, something like this:

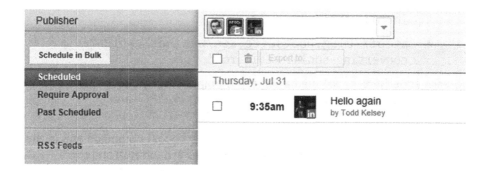

ROI Strategies

Hootsuite includes analytics and advanced monitoring capabilities. You can use Hootsuite to measure return on investment, but that ultimately depends on how you measure ROI. I certainly think it is worth learning about, even though it seems like ROI can be somewhat fuzzy.

Personally, I think it's pretty solid to say that Hootsuite and ROI most likely connect when you develop a regular stream of content about your business or organization that delivers relevant information about the products, activities, or heritage—essentially the "story" of your organization.

This invaluable content, which can also be posted on an official web site or blog, not only helps with search engine optimization (getting ranked well in Google search results), but it also helps people coming to your site. It's perfect content for business. The ROI is building a base of information about the company, the search engine optimization value. (It can help sales if you're selling something, so you can tie SEO to profit.) The content is rock solid, and Hootsuite helps you share it on your social networks.

However, there are some other ways to look at ROI, and I recommend taking a few minutes to look through these blog posts. They will give you a sense of the kinds of things people are doing, and the language they use around ROI and social media:

- `http://blog.hootsuite.com/build-and-measure-social-media-roi/`

- `http://blog.hootsuite.com/webtrends/`

- `http://blog.hootsuite.com/measure-social-media-roi-business/`

- `http://blog.hootsuite.com/8-tips-for-social-business-7/`

- `http://blog.hootsuite.com/social-media-roi-introduction/`

- `http://blog.hootsuite.com/webinar-content-converting-social-media-roi/`

To earn superhero social media awesome points, I *highly* recommend getting into the habit of doing Google searches on topics like "ROI and _____" (you fill in the social network or tool).

If you will forgive one *Star Wars* reference in this book, I command you in the name of Obi Wan Kenobi to try Google searches around ROI and tools. Read the articles and blog posts. People share their experiences, principles, and best practices.

(And, nudge nudge, you can blog about this too! The more exploration of ROI, the better, even if you're just writing about what you don't know. Even that can be helpful and encouraging to other people, who are out there doing the same kind of Google searches.)

Learning More

Congratulations on making it almost all the way through the chapter! Remember that there are some good resources out there for learning about Hootsuite and related topics.

■ **Tip** For some thoughts on how often to post, see http://www.nimble.com/blog/posting-and-analyzing-on-facebook/ and http://blog.bufferapp.com/social-media-frequency-guide (you can also try doing a Google search on the topic of how often to post on social media).

Conclusion

Thanks for reading this chapter. I think Hootsuite is a good tool for social media marketers to use—it's an essential item in your toolbox. As helpful as it is, it doesn't create the content for you, so in some ways, I think that Chapter 2 is the most important chapter of all. But anything that makes it easier for you to work with content frees up your time to *develop* the content.

Social Media Monitoring and Analytics

This chapter looks at the practice of social media monitoring, which is also known as *social listening*. It's an important part of social media marketing. We'll also take a look at *social analytics*, which can help you see how your social media efforts are going. We'll explore several related tools and try things out.

Understanding Social Media Monitoring, Social Listening, and Analytics

A large part of social media marketing is initiating conversations and putting content on social networks, but another significant part of it is listening to what people are saying. For example, over time, one way to see how effective your efforts are is to see what people are saying about them—either through direct responses or on their own social media accounts. In theory, you could read through every post, but how do you find them? What do you do if there are a lot of posts? Tools have evolved to help you see how things are going.

© Todd Kelsey 2017
T. Kelsey, *Introduction to Social Media Marketing*, DOI 10.1007/978-1-4842-2854-8_8

At a basic level, social analytics similarly allows you to measure impact, such as the number of likes, or conversations, or the performance of a particular page or account. This can be helpful for reporting to a client or colleague about how your social media efforts are going, and it can also help you effectively allocate your time and resources.

What Is Social Media Monitoring? Social Listening?

In general, social media monitoring and listening are two terms for the same thing. They can be divided into *owned media* and *earned media*.

Owned media includes social media pages or "properties" that you control, such as your Facebook page, Twitter page, and blog. These are the places where you post content and start conversations, and where you have direct access to reviewing the conversations. Technically, you could just look at what's going on directly—this is more realistic when a social media property is just starting—but the more accounts you have, the more conversations go on, the more likely you'll need tools that help you deal with this greater volume of conversations. These social media monitoring tools attempt to make your life easier, and they help provide insight into any trends.

On the other hand, *earned media* is everything else. That is, when people start their own conversations, or maybe someone shares something you post with someone else. The Holy Grail of social media marketing is a self-propelled, positive word of mouth conversation with relevant content, offers, or news. You could also think of earned media as "off the ranch," because it is out there, not on a page you control, but anywhere other than that page. So how do you find out what people are saying? There's an app for that. This is another area where social media monitoring is trying to make your life easier.

Tools

To get a sense of the space, here are a few excerpts from recent job descriptions on LinkedIn related to social media monitoring:

- Listening, monitoring, and measurement: Time allocation 20%
    - Be the eyes and ears of the brand(s) across the entire social space
    - Identify *conversations or crises* that require response and route through the appropriate department for resolution
    - Establish and use *listening tools* to gauge the health of the brand(s) online, and potential for participating in new communities

- Partner with the *analytics* team to ensure accurate tracking of social media initiatives

- Provide off-hours moderation of all brand(s)' social media channels

- Pull *reporting* and provide input on weekly, monthly, and ad hoc social media dashboards and reports

- Experience using *social media listening* and analytics tools such as *Facebook Insights, Twitter Analytics,* and Sysomos

- Digital analytics tools (Omniture, Google Analytics, *Facebook Insights, Radian6*) a plus

Note the terms in italics. If you're learning the space, one way to learn about tools and related skills is to scan job descriptions. Sometimes you can scan for product names (for example, Radian6 is a top social media monitoring tool), and other times you can scan for skills, such as "social monitoring" or "social listening".

Like other social media marketing skills, social monitoring in some cases can be a dedicated role, but it can also be one skill in a "multiple hats" role. This is when it's part of your skillset where you are posting content as well as listening to what's going on.

The Need for Speed: What Is a Social Media Crisis?

As I look out in the space of social media marketing, I believe that monitoring for crisis situations can be one of the most cost-effective ways to use social media. (This is in addition to thinking about your return on investment to track if and when social actually generates revenue.)

In some cases, it's a blind spot. Until you have a public relations crisis, it's hard to get a clear sense of how costly one can be. But it can be a really big deal and can cause significant financial damage for any business. It can cost millions of dollars to "clean up" a public relations crisis, and for smaller businesses, while the dollar amounts may not be as great, it can make the difference between being profitable or not.

The main reason it's a really important issue in social media marketing is because social media allows messages to spread so quickly. I remember being at a conference about social media. A rep from a big PR firm was there talking about PR crises, and how the time you have to respond to a crisis before it gets out of control has reduced. At this point, it's arguable that it can be a matter of hours before something bad gets national media attention.

For example, consider what happened to Target in 2013. It had a "data breach" (which, by the way, can happen to *any* business).

In "Five Lessons for Every Business from Target's Data Breach," Natalie Burg argues that companies need to be very proactive:

1. Communicate the problem, pronto

    "The company moved quite slowly on this breach," reported John Biggs for TechCrunch on the same day as Target's official announcement regarding the crisis. That's because a report on the breach had emerged a week earlier by Krebs on Security. Though just rumors at that point, the post turned out to be so accurate—down to how the theft likely occurred—that it revealed how long Target likely knew a crisis had occurred before alerting customers. On the other hand, they could have been even slower to respond. As the recent security breach announcement by Snapchat has revealed, waiting months to acknowledge and even longer to apologize for a breach can amplify the problem.

    ```
    http://www.forbes.com/sites/sungardas/
    2014/01/17/five-lessons-for-every-business-
    from-targets-data-breach/
    ```

Another article by Ronn Torossian and titled "How to Use Social Media to Improve Crisis Communications," shows a responsive strategy:

> *For example, when a McDonald's worker threatened and verbally attacked a customer on camera, the company responded to the victim privately and apologized, while also assuring the public that they were aware of the issue and had fired the employee.*

```
https://www.forbesw.com/sites/
forbesagencycouncil/2016/09/09/how-to-use-
social-media-to-improve-crisis-communicatio
ns/#3d9d17912f16
```

Naturally, it takes resources to monitor things on social media, but when you consider the potential fallout, it's a good idea to be monitoring things. Some places recommend having a simple social media crisis plan, which can also serve the purpose of showing how costly things can be if you don't respond quickly enough.

On a smaller scale, social media monitoring around potential crisis situations can also be a long-term, strong leg to stand on. Although social media in this case isn't necessarily generating money per se, it's a safeguard, insurance of sorts, against losing a significant amount of money or good standing with customers.

I suggest all corporation and organizations look into it. If you're a social media marketer, it should be on your list of services you provide.

Social Listening in Hootsuite: Owned Media

Hootsuite is a simple tool we looked at in the previous chapter, which also allows you to do social listening on owned media in a central place. At some point Hootsuite may expand its offering to earned media as well.

To take a look at the options, you can log in and look at the streams that you set up. For example, if your Twitter streams are set up, Hootsuite will give you a feed of what people are saying on Twitter and indicate when you have new followers.

For example, one tactic on Twitter for growing your number of followers is to follow the people following you. In some cases, you might see how many followers they have, and if they appear to be influential, you might engage them directly in conversation. Looking at owned media can be a way of seeing what's going on as well as building a strong following.

Basic Social Analytics

Social analytics can be a focus area or even a dedicated role. There has been increasing interest in it, and new tools coming out, because there is a trend toward companies wanting to track the ROI of their social media efforts (surprise). In some cases, this means measuring trends, such as an increase and growth in social media properties, or determining if the conversation around a particular company or campaign is positive or negative.

One of the simplest ways to see social analytics in action is to go to a Facebook page you created and click on the Insights link:

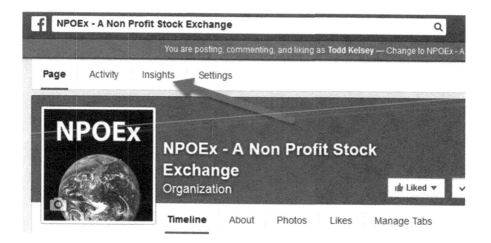

There are ways to get "deep data" about social media activity, but the basic interface provides information that anyone can look understand.

When you click on the People link, you can get insights about the people who are visiting your page:

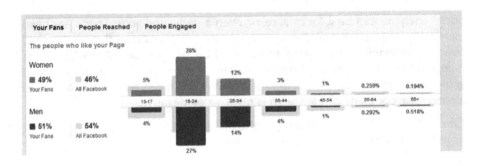

For example, one thing that was really interesting about making an international Facebook page during the Arab Spring was to see which countries people were visiting from:

Overview	Likes	Reach	Visits	P
Egypt			1,764	
Bahrain			623	
Yemen			258	
United States of America			65	
Saudi Arabia			52	
India			32	
Philippines			24	
United Kingdom			19	
United Arab Emirates			18	
Tunisia			16	

Basic Social Reporting in Hootsuite

When you start looking at these types of things, especially if you're offering a paid service to a client or are reporting your progress to someone at work, social reporting can make your life easier. You could go to all the separate tools, but sometimes it can be nice to have things together in a central place.

Hootsuite allows you to do some social reporting, including scheduled, automatic reporting that delivers an e-mail to your inbox.

To try it out, roll over the toolbar on the left side in Hootsuite:

Click on Analytics:

There are a variety of reports and templates you can use.

To get started, just click Create One Now:

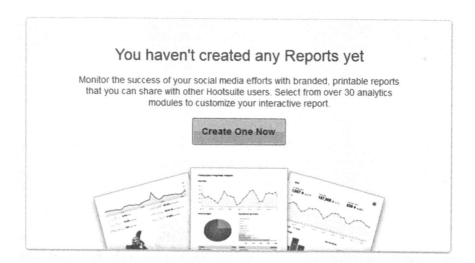

Then select a template. Some of them are only available in the paid version.

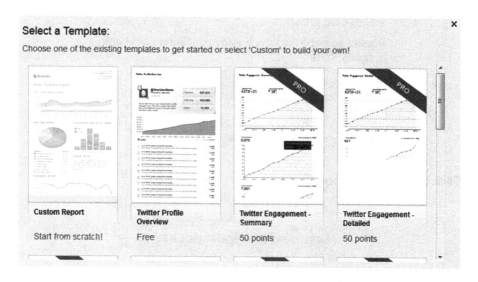

Scroll down and find the free one called Facebook Page overview, which you can try. (This example assumes you've created a Facebook page and integrated it into Hootsuite.)

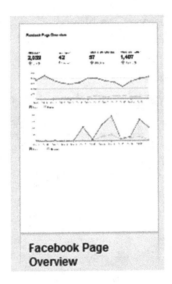

**Facebook Page
Overview**

You can click the template to select it. Then you can select the Facebook page that you want to generate a report on:

Click Continue to Report Builder:

Continue to Report Builder

Next you'll end up on a screen you can explore, with a report template:

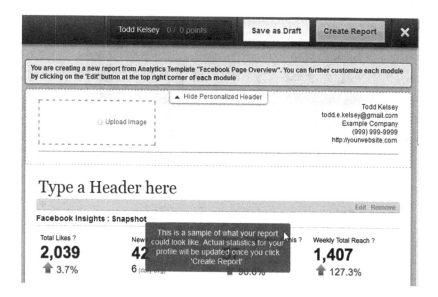

You can also just click Create Report:

The report will be filled in with any data you have:

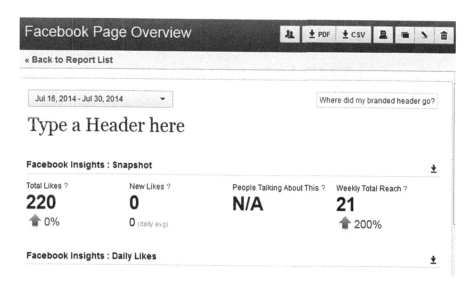

If you haven't typed in a name for the report, you can always click the Edit button:

Type a header:

Then click Save Changes:

You'll have a nice, nifty custom report. Such a report includes information you can get in Facebook insights, but it makes it easier to share with other people.

When you're looking at these types of reports, try to set the different date ranges. For example, you might generate a report once a month, or once a week, and review it with colleagues or with your clients. You can click on the date range drop-down menu and choose an option:

You can also try different templates:

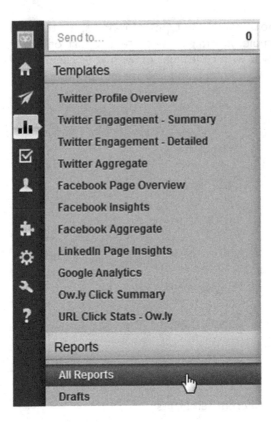

My general recommendation is to focus on your social media content and campaign first. When things are going along nicely, work on building your skills in being able to report on the activity.

Hootsuite allows you to come back and access reports you've made at any time:

Scheduling and Automating Reports

Scheduling and automating a report can be a nice way to save time and have a tool for providing regular insight.

To schedule a report, in the report, click Edit:

A drop-down menu appears where you can choose the frequency:

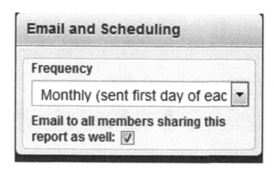

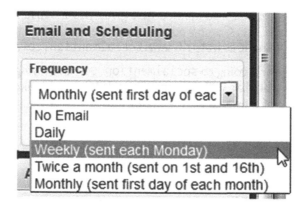

In order to actually set it in motion, you need to click on Share:

Then you add e-mail addresses.

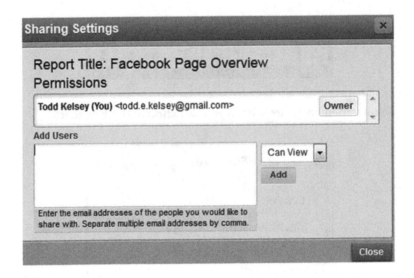

Trying Socialmention.com

Socialmention.com is a free service that gives you a sense of what social media monitoring is like. Compared to a tool like Radian6, it has limitations, but it's free. You'll find it at http://www.socialmention.com/.

I recommend regularly going to socialmention.com and trying various kinds of searches to get familiar with its capabilities. It looks a bit like Google.com, and you can think of it as a social search engine:

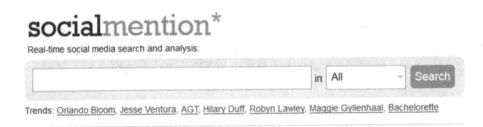

You can type in a brand name, celebrity, company name, or topic:

The All drop-down menu allows you to select social media channels. At first, you might just choose All and then click Search:

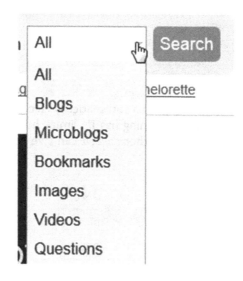

You'll see a variety of figures. Don't be alarmed!

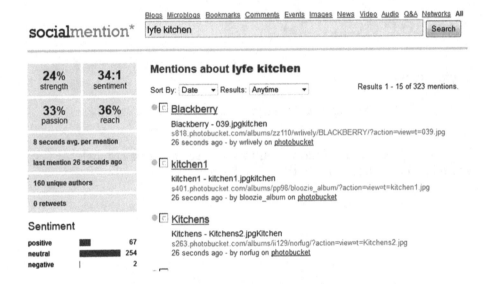

On the left, social mention attempts to automatically "score" the general social perspective. Automated social listening has its limits, but depending on how many conversations or "mentions" there are, it can give you a high-level view:

24% strength	34:1 sentiment
33% passion	36% reach

8 seconds avg. per mention

last mention 26 seconds ago

160 unique authors

0 retweets

To learn more about how social mention works, see `http://socialmention.com/faq`.

Sentiment is an automated attempt to see if people are speaking positively or negatively about a topic.

Sentiment

positive	▬	67
neutral	▬▬▬▬	254
negative	|	2

The Value of Quotes

The Top Keywords list shows you some of the keywords related to your search, but it can also show a blind spot. In typing "lyfe kitchen" in my original search, I didn't use quotes around the phrase, so it was looking at the words separately, including mentions of just "kitchen". (I wanted lyfe kitchen as a brand name, which is a new restaurant chain.)

Top Keywords

kitchen	▬▬▬▬	481
lyfe	▬▬▬	269
food	▬	107
restaurant	▮	46
good	▮	46
dont	▮	44
time	▮	42
link	▮	33
people	▮	32
chef	▮	31

The moral of the story is, when searching on this tool, try using quotes around exact phrases:

Blogs Microblogs Bookmarks Comments Events Images News Video Audio Q&A Networks All

| "lyfe kitchen" | Search |

The results may turn out differently. In this case, it is a more focused query:

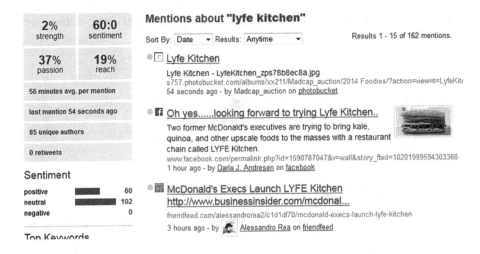

The sentiment is a bit different:

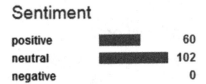

Other areas to explore include Top Hashtags. A post on your search may fall into such a hashtag:

Top Hashtags

yelp		3
tytmeetups		1
tytlive		1
biteflite		1
whatsforlunch		1
kitchen		1
smarp		1
design		1
lyfekitchen		1
14yoj		1

See Chapter 5 on Twitter as well as `http://en.wikipedia.org/wiki/Hashtag` for more information on hashtags.

The Sources area is also interesting. It mentions the many channels that posts can appear on:

Sources

friendfeed		82
youtube		49
reddit		9
topix		9
ask		4
wordpress		4
Yahoo News		3
facebook		1
photobucket		1

Another thing you can do is look at a single channel. For example, click the Microblogs link at the top:

You will see "only" the results in microblogs, such as Twitter:

To get a sense of how social monitoring works on an ongoing basis, try using the Email Alert function on the right side of the screen:

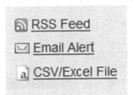

Other Tools

There are many social monitoring tools. Radian6 is a top commercial tool, but there are many good competitors, and the field is always evolving. As an exercise, I recommend exploring the following searches on Google:

- Top social media monitoring tools
- Top free social monitoring tools
- Top social listening tools

ROI Strategies

Social monitoring is on the fuzzy side of social media, in that there aren't many tools that combine a true way to figure out the financial impact of engagement or sentiment. This is partly because unless a social monitoring or analytics tool is tied directly to ecommerce, where you can trace the financial impact, it's hard or impossible to measure the ROI.

But for purposes such as monitoring for potential crises, and to get a sense of what people are talking about, it is pretty helpful. Some companies spend most of their efforts listening for conversations in which someone is complaining about a product or has a question, so that customer service can respond quickly and directly on social media.

My general recommendation when considering leveraging social monitoring for an organization or client is to focus on crisis monitoring. Make a crisis plan and do some ongoing monitoring for customer service issues (no business is too small for this, even locally run businesses). Social monitoring can also be used to get a sense of how people react to a new campaign.

Building a following on Facebook or Twitter can be a valuable sounding board for testing new ideas and campaigns. You may get a comment or conversation going, and you get valuable feedback in the process.

Social analytics are closer to being able to track revenue. If you learn about web analytics tools such as Google Analytics or Omniture, explore tools like Argyle Social, and invest some time figuring out how to share coupons or offers on social media, you can look at how people share or click on an offer you post and maybe even trace it back to actual sales. For most businesses, though, the value of social monitoring is probably strongest for crisis monitoring and customer service.

But hey, challenge me! (on the Digital Marketing Cafe LinkedIn group).

Hootsuite Certification

To hone your skills and to have something to put on your resume, you might be interested in exploring Hootsuite Academy, which has some additional learning material. At present, there's a cost to it, but the certification is doable and can be something nice to work toward:

To explore it, visit `https://hootsuite.com/education`. Then find the Hootsuite Platform Training area and click Get Started:

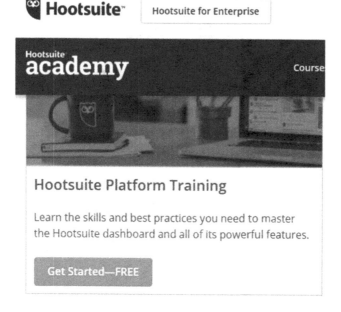

The courses themselves are free:

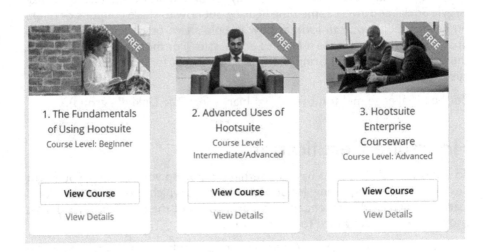

When you want to get certified, you select the certification:

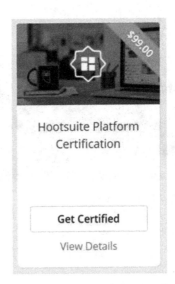

At the time of this writing, it's $99, which I think is worth it to have as a credential. Ultimately, you need to get experience and learn to use the tool that matters, but certification can be a nice way to help you develop momentum. It's also looks good on your resume.

OMCP/Market Motive Certification

Another certification and set of courses on social media in general that you might like to explore is OMCP certification. It's supported by a company called Market Motive, which you can see at http://www.marketmotive.com/.

There are a variety of courses, and you can earn "badges," so it's nice to work toward. It may provide a boost to your resume or LinkedIn profile.

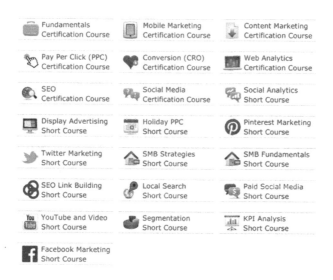

Free Radian6 Training

Until/unless Radian6 develops a trial version, it's a bit of a chicken and egg situation. Radian6 is a good thing to have on your resume, but it is a more expensive too. However, I definitely, strongly, recommend watching some Radian6 videos, to get a sense of how one of the leaders handles social monitoring. See https://www.youtube.com/playlist?list=PLWzH8Vzi6r wawin6sf5ehC2LPUZ7ijWA2 or search on YouTube for Radian6 training.

Learning More

In addition to the previous links and searches, here are a few items that are worth exploring:

- http://www.brandwatch.com/2013/08/top-10-free-social-media-monitoring-tools/

- http://analytics.twitter.com

- https://support.twitter.com/articles/20170934-twitter-card-analytics-dashboard

Twitter analytics is more of a social advertising analytics platform, but it's still worth knowing about, and possibly even trying. YouTube has an analytics section. Try Googling and add them to your LinkedIn profile and resume.

Conclusion

Congratulations on making it through the chapter and through this book! Social monitoring can be interesting; I think the best way to approach it is to try a tool like Hootsuite, use socialmention.com, and watch some Radian6 videos. Then concentrate on developing good content. Once your social media presence is going, I recommend coming back and actually using the tools. Best wishes!

■ **Special Request** Thank you for reading this book. If you purchased this book online, please consider going on where you purchased it and leaving a review. Thanks!

I

Index

A, B

Blog
  advantage of, 22
  post, 16
  search drill, 21
Blogger.com
  choose tool, 17
  create, 17
  settings menu, 20
  sign in, 17–18
  template, 19
Boosted posts, 37, 39–40
Business to business (B2B)
  channels, 9
  LinkedIn, 87–88, 94
  longer sales cycle, 4
  research, 4
Business to consumer (B2C)
  Amazon, 4
  LinkedIn, 88
  setting, 9
  Walmart, 4

C

CAPTCHA code, 25
Classic version, 23
Click-through rate (CTR), 58
Collaboration
  vs. creation vs. curation, 15–17
  defined, 16
Content post, 8

Conversion, 74
Creation
  vs. curation vs. collaboration, 15–17
  defined, 16
Curation
  vs. creation vs. collaboration, 15–17
  defined, 15

D

Digital images, 30
Digital marketing, 3

E

Earned media, 124
E-mail address, 92
Engagement, 5

F

Facebook ads
  budget, 64
  campaign
    create, 59–60, 62–64
    monitoring, 71–72
  free, 73
  headline, 68–69
  lifetime budget, 65–66
  metrics, 57–58
  overview, 53–55
  payment, 64
  purchase intent, 57
  targeting, 58–59
  uses, 56

© Todd Kelsey 2017
T. Kelsey, *Introduction to Social Media Marketing*, DOI 10.1007/978-1-4842-2854-8

Get the eBook for only $5!

Why limit yourself?

With most of our titles available in both PDF and ePUB format, you can access your content wherever and however you wish—on your PC, phone, tablet, or reader.

Since you've purchased this print book, we are happy to offer you the eBook for just $5.

To learn more, go to http://www.apress.com/companion or contact support@apress.com.

Apress®

Printed in the United States
By Bookmasters